Tank Irrigation of Dry Zones in India

About the Author

Dr. Sebak Kumar Jana is currently Associate Professor in Economics and Head of the Department of Economics with Rural Development, Vidyasagar University, West Bengal, India. Dr. Jana was graduated from Presidency College, Kolkata in 1989 and obtained M.Sc. degree in Economics from University of Calcutta in 1991. He obtained M.Phil. and Ph.D. degree in Economics from Jadavpur University, Kolkata. His area of research includes environmental and resource economics and economics of education. He has publications of about thirty peer reviewed research papers in national and international Journals and in edited volumes. He has completed research projects sponsored by UGC and ICSSR. His books include "Industrial Pollution and Abatement – Economic Analysis of Paper Units in West Bengal" (2010, South Asian Publishers Private Limited, New Delhi, Co-authored), "Education in West Bengal – Looking Beyond Schools" (2012, Shipra Publications, New Delhi), "Development in Developing Economies" (2012, Concept Publishing Co. Pvt. Ltd., New Delhi, Co-edited).

Tank Irrigation of Dry Zones in India

A Sustainable Livelihood

Sebak Kumar Jana

CONCEPT PUBLISHING COMPANY PVT. LTD.
NEW DELHI-110059

ISBN-13 : 978-93-5125-091-3

First Published 2014

Published and Printed by

Concept Since 1974

Concept Publishing Company Pvt. Ltd.
A/15-16, Commercial Block, Mohan Garden
New Delhi-110059 (India)
T: 25351460, 25351794, *F*: 091-11-25357109
E: publishing@conceptpub.com, W: www.conceptpub.com
Editorial Office: H-13, Bali Nagar, New Delhi-110 015, India.

Cataloging in Publication Data—*Courtesy:* D.K. Agencies (P) Ltd. <docinfo@dkagencies.com>

Jana, Sebak Kumar.
Tank irrigation of dry zones in India : a sustainable livelihood / Sebak Kumar Jana.
p. cm.
Includes bibliographical references (p.) and index.
ISBN 9789351250913

1. Tank irrigation—India—West Bengal. 2. Arid regions agriculture—India—West Bengal. 3. Irrigation farming—India—West Bengal. I. Title.

DDC 631.587095414 23

Foreword

Historically the small scale water resources had played a significant role in rural life and particularly in the irrigated agriculture. Irrigation tanks in India are century old water harvesting structures. There are about 208,000 tanks in the country irrigating about 2.3 m.ha. Marginal and small farmers and landless households are more benefited from tank multiple uses. However, encroachment, poor management, privatization and government appropriation of the tanks have been the main outcomes of the failure of local authority system to enforce the institutional arrangements under common property resources management regime. Poor management of the tanks had also resulted in lower land productivity compared to canals and wells. In addition to the existing constraints, the emerging threats in the form of outmigration of people from tank based agriculture also affect the stability of tank irrigation. However, due to constraints in the future expansion of the canal and well irrigated areas, tank irrigation still offers more scope for revival. Several interventions such as revenue mobilization, augmenting groundwater resources in the tanks, strengthening the multiple uses, tank sluice management, integrating social forestry and desilting, and tank modernization are being examined in different states.

In Eastern India, these numerous surface water bodies are under the various management regimes that include tanks being governed by Panchayats or state government as minor irrigation tanks or independent tanks owned by individuals/groups. Some ponds were built by landlords, and many remain under private ownership today, while others were built by public institutions such as temples, that continue to manage them today. Unlike the South India tanks which are mainly used for irrigation, a major number of government tanks in East India are used for fisheries. Multiple use wetlands in the regions mainly include: water for irrigation, water for domestic use (direct consumptive use), jute retting (in-stream use), cultivation in the pond bed (direct use), fisheries (direct use), and recharging groundwater (indirect use). The holistic contribution of the tanks in the region needs to be studied in detail. In the recent years, groundwater has emerged as a supplemental source to refill tanks depending on location of the tanks/ponds but availability of electricity and diesel, and the investment necessary to bore wells constrain their full utilization. Thus, the importance of the small scale water bodies is decreasing over years due to subsequent droughts, poor water control and poor management. Across eastern India, the highest per cent decline in number of tanks over a period of 14 years is found in West Bengal with a 92 per cent decline, closely followed by Uttar Pradesh (76%), Odisha (59%) followed by West Bengal (55.6 %).

While improving smallholder irrigation systems in East India, the issue is considered as one of increasing 'access' to water supplies and technology adoption and 'management'. Improving the availability of tank/pond water supplies will minimize the burden on groundwater pumping due to poor electricity supplies and costly diesel pumping. By conjunctive use of the surface and groundwater supplies along with the enhanced multiple uses will improve the livelihood of the households.

As such limited publications are available on tanks in eastern India and several research studies are also underway focusing on the livelihood opportunities related to multiple uses of small scale water bodies in eastern India. In this context, the research works undertaken by Dr Sebak Jana are timely and in a very short period he is able to compile his research publication in the form of this book. The book has interesting chapters focusing on the basic characteristics of the tanks to management of the tanks for enhancing the multiple use benefits.

I am sure this book will be very useful to researchers, NGOs, planners and policy-makers dealing with small scale irrigation tanks in India and elsewhere.

I congratulate Dr. Jana for bringing out this very valuable book at a time when the tank irrigation is getting the attention of the Government departments.

Hyderabad
3, January 2014.

K. Palanisami
Principal Researcher
International Water Management Institute (IWMI)

Preface

Rainfed areas in India still account for about 60 per cent of the cultivated area and are home to the majority of rural poor and marginal farmers. The livelihood of rural people has often been severely affected by repeated droughts and erratic rainfall. The irrigation sector will also likely be strongly affected by climate change in India with the predicted increased variability in precipitation. It is well documented in the literature that irrigation plays a vital role in improving the yield or productivity in the agricultural sector and the impact of irrigation on reducing poverty is striking. Among the irrigation sources, tank irrigation is considered important in the dry zones as it can play a very critical role in the sustainable irrigation development in these areas. Tank generally means a small scale surface reservoir deriving the inflows mainly from rainfall run-off. Tank irrigation has various benefits, namely tanks can have a wider geological distribution than large-scale projects, tank investments tend to be less capital intensive, tanks have fewer negative environmental impacts and tanks can provide a variety of livelihood options to the rural people. In spite of these advantages, tank irrigation has declined in almost all states in India over the years. One of the important causes of a decline in tank irrigation is weakening community co-operation. The potential importance of tank irrigation can also be taken from the fact that ground water is over-exploited in some places in India leading to the threat of water scarcity. The present book highlights the problems and prospects of tank irrigation in the zones in West Bengal. The study is based on the primary survey carried out on 181 tanks in the three districts of West Bengal.

In Chapter 1 we have examined the pattern of irrigation

development in West Bengal. We have seen the district-wise percentage of cultivable irrigated area has a wide disparity among districts. A literature review on tank irrigation has been presented thematically. The objectives, methodology and study area are also given in the chapter.

In Chapter 2, we have mainly dealt with agro-climatic characteristics of the sample districts, namely Bankura, Purulia and Paschim Medinipur. The analysis has been done in comparative framework. The source-wise irrigation development for all the three districts for the period 1996-97 to 2005-06 has been documented. Block-wise tank irrigated area and change for the two sub-periods (1996-2000) and (2001-2005) have also been compared.

In Chapter 3, we have presented findings of the tank characteristics and performance of the tanks under study based on primary survey. The characteristics include – extent of irrigation, conditions of tank and irrigation efficiency.

In Chapter 4, we have analysed the WTP behaviour of the households for the tank renovation/reconstruction with Tobit Analysis. We have used factor analysis (Principal Component Analysis) for tank quality assessments. Tank quality has been judged on 14 variables. Tobit regression analysis has been done to find the determinants of cropping intensity. We have used cost benefit analysis to judge the economic viability of *hapa* (a small private tank). The watershed management programme has strong impact on the livelihoods of rural people. These programmes have tremendous economic impact for sustainable livelihood generation. In the last chapter we have given some suggestions for improving the conditions of tanks.

The book is based on ICSSR sponsored Research Project entitled "Prospect of Sustainable Tank Irrigation Option in the Dry Zones of West Bengal". I must put forward my gratitude and sincere thanks to ICSSR authority for the kind approval and financial assistance of the project. I must also convey my sincere thanks to the anonymous consultant for ICSSR for valuable commens and suggestions made on the Report.

I must pay my sincere thanks to Professor Ranjan Chakrabarti, Honourable Vice Chancellor, Vidyasagar University for his kind inspiration to publish the Research Project Report as a book.

The research assistants of the project, namely Mr. Amit Das and Mr. Bikash Dinda have tried hard in data collection and analysis of data. Mr. Sandip Barman and Mr. Siddhartha Sankar Manna have helped us a lot in the data processing of the project.

I am also deeply indebted to the faculty of the Department of Economics with Rural Development, Vidyasasagar University to which I belong for extending every co-operation which I required. I am also thankful to Professor Joyshree Roy and Dr. Asim K. Karmakar of Department of Economics, Jadavpur University, Kolkata for their kind advice.

I gratefully acknowledge the assistance extended to us by several officials at various levels like District Irrigation Office, District Principal Agricultural Office, District Planning Office, Zilla Parishad, Soil Conservation Department, Block Development Office, Gram Panchayat Office in our selected districts of Paschim Medinipur, Bankura and Purulia in West Bengal. I am also indebted to various officials at State Water Investigation and Development Department for their advice relating to various aspects of the project. Also I have received great help from the Officials of PRADAN in relation to the project.

No amount of thanks is sufficient for those persons who have spared their time in responding to our questionnaire for the primary survey. It was difficult for me to prepare a complete list of the persons to whom I am indebted for the project directly or indirectly.

I must mention specially the name of Dr. K. Palanisami of IWMI, Andhra Pradesh for his valuable observations, suggestions and comments on the work.

The secondary data have been accessed through different sources. I am indebted to the following Libraries—Vidyasagar University Library, Midnapore, Presidency College, Kolkata, ISI, Kolkata, National Library, Kolkata, Bureau of Applied Economics and Statistics, Kolkata.

Last but not the least the role of my family members for completion of the project work can in no way be neglected.

Our humble endeavour will be successful if it stimulates further interest among future researchers to take up unanswered questions in this book.

Vidyasagar University
Midnapore

Sebak Kumar Jana

Contents

List of Tables

List of Figures

List of Map

List of Plates

Abbreviations

AIBP	Accelerated Irrigation Benefits Programme
ANOVA	Analysis of Variance
APL	Above Poverty Line
BCR	Benefit Cost Ratio
BDO	Block Development Officer
BPL	Below Poverty Line
BRGF	Backward Region Grant Fund
CAD	Command Area Development
CADA	Command Area Development Authorities
CADWM	Command Area Development and Water Management Programme
CCA	Culturable Command Area
CDI	Crop Diversity Index
CRS	Constant Return to Scale
CVM	Contingent Valuation Method
CWDP	Common Watershed Development Project
DDP	Desert Development Programme
DEA	Data Envelop Analysis
DMU	Decision Making Unit
DPAP	Drought Prone Areas Programme
DRDA	District Rural Development Agency
DRS	Decreasing Return to Scale
DWCRA	Development of Women and Children in Rural Areas
EAS	Employment Assurance Scheme
EPW	Economic & Political Weekly
FPARP	Farmer's Participatory Action Research Programme
Ft/ft	Foot

GDP	Gross Domestic Product
GWL	Ground Water Level
GOI	Government of India
GOWB	Government of West Bengal
GP	Gram Panchayat
GS	Gram Sansad
GUS	Gram Unnayan Samiti
Ha./Hec.	Hectare
HDTW	High Capacity Deep Tubewell
INRM	Integrated Natural Resource Management
IPCC	Inter Governmental Panel on Climate Change
IRDP	Integrated Rural Development Programme
ITDP	Integrated Tribal Development Programme
IWDP	Integrated Waterlands Development Programme
JGSY	Jawahar Gram Samridhi Yojana
JRY	Jawahar Rojgar Yojana
KBK	Koraput, Bolangir & Kalahandi
KMO	Keiser Meyer Olkin Criteria
LDTW	Low Capacity Deep Tubewell
mbgl	meter below ground level
MDTW	Middle Capacity Deep Tubewell
Mham	Million hectare meter
MOWR	Ministry of Water Resources
Mt/mt	Metric Tonne
MWS	Million Well Scheme
muni	Municipality
NABARD	National Bank for Agriculture and Rural Development
NBTDP	North Bengal Tarai Development Project
NGO	Non-Governmental Organisation
NPV	Net Present Value
NREGA	National Rural Employment Guarantee Act
NREGS	National Rural Employment Guarantee Scheme
NREP	National Rural Employment Programme
NSSO	National Sample Survey Organisation

NWDPRA	National Watershed Development Project for Rainfed Areas
O & M	Operation and Maintenance
OBC	Other Backward Caste
ODW	Open Dug Well
PFA	Project Facilitating Agency
PIM	Participatory Irrigation Management
PM	Paschim Medinipur
PPP	Public Private Partnership
PRI	Panchayati Raj Institution
PWD	Public Works Department
Qntl/Qt	Quintal
RKVY	Rashtriya Krishi Vikas Yojana
RLEGP	Rural Labour Employment Guarantee Programme
RLI	River Lift Irrigation
RRR	Repair, Renovation and Restoration
RSVY	Rashtriya Sama Vikas Yojana
SAE	Sub Assistant Engineer
SC	Scheduled Caste
SGRY	Sampoorna Grameen Rozgar Yojana
SGSY	Swarnajayanti Gram Swarozgar Yojana
SHG	Self Help Group
ST	Scheduled Tribe
STW	Shallow Tubewell
SWIDD	State Water Investigation & Development Department
TEV	Total Economic Value
TRYSEM	Training of Rural Youth for Self Employment
VDC	Village Development Committee
WB	West Bengal
WDF	Watershed Development Fund
WDP	Watershed Development Programmes
WIDD	Water Investigation and Development Department
WM	Water Management
WRDD	Water Resource Development Directorate
WTP	Willingness to Pay
WUA	Water Users' Association

Abbreviations Used in the Analysis of Data

Abbreviations	*Full Name*	*Units*
ADEWATER	Adequacy of tank water	Dummy
AGEHED	Age of the head	Year
AGLAND	Agricultural Land	Dummy
AREA	Area of the Tank	Acre
ASSOCIATION	Whether there is association	Dummy
BOROD	Boro Dummy	Dummy
CERTAINT	Certainty of getting water	Dummy
COMMUNIC	Communication score in the village with scale 1 to 5	Dummy
CONTWAT	Control of water use	Dummy
CROPSHOC	Crop failure	Dummy
DEPTH	Depth of the tank	In Foot
DISTPAKK	Distance from Pakka Road	Kilometer
DCOMMUNITY	Community Owned Tank	Dummy
DGOV	Government Owned Tank	Dummy
DOANSD	Quality of soil is loamy	Dummy
EDUHEAD	Education of Head of the Family	Percentage
EDUHEAD	Education of Head of the Family	Unit Free
EFFSCORE	Efficiency scores	Unit Free
ENTRYEXI	Water Entry and Exit System	Dummy
EXPINP	Expenditure on Input	Rs.
EXTEN	Whether the household gets facilities of extension activities	Dummy
FARFARME	Far farmer gets less water	Dummy
FARMOUTP	Farm Output	Rs.
FEMADULT	Female Adult	Number
FISHERY	Whether there is fishery in the tank	Dummy
FOREST	Afforestation	Dummy
FUNDAV	Fund availability	Dummy
GALAND	Gross agricultural land possessed by the household	Acre

GCA	Gross command area of the tank	Acre
IPBNIP	Irrigated price by non-irrigated price in the tank command area	Rs.
LINING	Lining between Channels and Land	Dummy
LOGFERTI	Log of fertilizer expenditure	Rs.
LOGGCA	Log of gross cropped area	Acre
LOGK	Log Capital	Rs.
LOGL	Log Labour	Rs.
LOGLAND	Log value of land	Rs.
LOGSEED	Log of seed expenditure	Rs.
LOGTANKC	Log of tank irrigation cost	Rs.
MAINBUND	Maintenance Bund of the tank	Dummy
MAINCH	Maintenance of Channels	Dummy
MAINTANK	Maintenance of tanks	Dummy
MALEADUL	Male Adult	Number
MALEWAGE	Male Wage Rate	Rs. per day
MUSTARDD	Mustard Dummy	Dummy
NOFFFARM	No. of off-farm workers	Number
OBC	Other Backwad Caste	Dummy
OTHERPROP	Proportion of crops other than Aman	Proportion
PAKKA	Whether there is pakka road in the village	Dummy
PANCHYAT	Role of Panchyat in tank development	Dummy
PANCHYAT	Panchayat Tax	Dummy
POTATOD	Potato Dummy	Dummy
PUMP	Whether there is pump	Dummy
RELATION	Relation among the members	Dummy
SEASMUMD	Seasmum Dummy	Dummy
SHAREBOR	Share of Boro	Percentage
SHAREMUS	Share of Mustard	Percentage
SHAREPOT	Share of Potato	Percentage
SHARESUN	Share of Sunflower	Percentage
SHAREVEG	Share of vegetable	Percentage
SHAREWHE	Share of Wheat	Percentage
SUEZ	Suez system	Dummy
SUNFLD	Sunflower Dummy	Dummy
TOTPODY	Productivity at tank level	(Rs./Acre)
VAL.FAT	Fertilizer expenditure for all crops	Rs.
VEGD	Vegetable Dummy	Dummy
WEALTHD	Wealth dummy	Dummy
WHEATD	Wheat Dummy	Dummy
WATERSHA	Water Sharing System	Dummy
YIELDAMA	Yield of Aman Paddy	Quintal/Acre

Chapter 1

Introduction

1.1 Introduction

1.1.1 Need for Irrigation

Indian economy even after six decades of independence is still an agrarian economy. Though the share of agriculture in GDP is about 14 per cent, 58 per cent of people in India are still dependent on agriculture for their employment and livelihood (*GoI, Economic Survey 2010-11*). It is to be pointed out that growth rate in agricultural GDP has decelerated from 3.5 per cent per year during (1981-82 to 1996-97) to about 2 per cent during (1997-98 to 2004-05) in India. Again where the GDP has grown at the rate of 8.6 per cent during 2004-05 to 2010-11, agricultural sector has increased by 3.46 per cent during the same period. India at present faces a daunting challenge to provide food security to the burgeoning population. For that, what is needed first is to enhance agricultural production. Agricultural production can be increased either through expansion of area or through yield improvement. As there is limited scope for area expansion, the opportunities for yield improvement should be explored. It is well documented in the literature that irrigation plays the vital role in improving the yield or productivity in the agricultural sector and the impact of irrigation on reducing poverty is striking. It is found that in most states households with access to irrigation have only about half the poverty incidence compared to the households without irrigated land. The effect of irrigation facility even among the tribal households is worth mentioning (*Roy 2006*). Rained areas in the country still accounts for 60 per cent of the cultivated area and are home to majority of rural poor and marginal farmers. Repeated droughts and erratic rainfall have severally affected the livelihoods of rural people.

Although this deceleration occurred in almost all states, it was most marked in the rainfed areas. This should be understood in global context also. Agriculture uses 85 per cent of water consumed in developing countries, mainly for irrigation. Irrigated farming accounts for 18 per cent of the cultivated area in developing countries and produces 40 per cent value of the agricultural produce. Many countries are facing worsening water scarcities. Globally 15-35 per cent of total water withdrawals are unsustainable – the use of water exceeds renewable supply. In all the regions poor will be disproportionately vulnerable because of their dependence on agriculture and their low capacity to adapt (*World Development Report 2008*).

In this backdrop, there is an urgent need to explore the possibilities of sustainable forms of irrigation. The problems with the current irrigation development strategy is heavy dependence on surface and ground water and secondly, dependence on the state (*Agarwal 2001*). The projected water demand in India is 105 mham in 2025 which was 38 mham in 1974. Out of this projected 105 mham water demand, 70 mham will come from surface water and the rest will come from ground water. This heavy reliance on river and ground water is creating a number of problems. In numerous rivers because of heavy exploitation, river flow in summer is getting dried. Construction of large dams and neglecting of small water harvesting structures are causing forced human displacement. The ground water table is falling in many parts of the country. Another problem with the current irrigation development strategy is that people are much dependent on the state for irrigation subsidy which is causing pressure on the state exchequer. So, there is an urgent need to explore a sustainable irrigation strategy. Tank irrigation can play a great role in this direction. A tank is a low earthen bund constructed across a shallow valley to hold the rainfall runoff from catchment area.

In this chapter we will discuss mainly various dimensions of tank irrigation development in West Bengal. Before going into the discussion of tank irrigation development in West Bengal, we will discuss irrigation development in West Bengal which will help to understand development of tank irrigation in the state.

1.1.2 Agriculture in West Bengal

West Bengal is the largest producer of rice and second largest producer of potatoes and a major producer of fruits, vegetables and oilseeds in India. However, the state is a net importer of pulses and oilseeds. The state has not achieved food security in any food crop except rice and potatoes. The net cropped area is almost close to 100 per cent of the cultivable land. And there is meagre possibility of increasing it. The alternative to increase production is to increase yield. This is possible because the yield of every important crop in West Bengal is less than respective leading states in the country, e.g. West Bengal's rice yield is 2.5 tons/he while it is about 3.8 tons/he in Punjab and Karnataka (*CII 2007*). Rice is the major crop in West Bengal. Major crop-wise area is presented in Table 1.1. Out of 83.95 lakh hectares of land under cultivation, total area under rice is 59.36 lakh hectares of land. The productivity of rice is 2.53 tons per hectare. Boro rice cultivation has been important factor in enhancement of agricultural production in West Bengal. However, the water intensity of the crop is a major limiting factor in the spread of boro cultivation in the dry zones like Bankura and Purulia where there is less rainfall.

Table 1.1: Area, Yield Rate and Production of Principal Crops in West Bengal, 2008-09

	Area (Lakh hec)	*% Area*	*Yield (Tons/hec)*	*Production (Lakh tons)*
Autumn (Aus) Rice	2.92	3.5	2.07	6.05
Winter (Aman) Rice	40.87	48.7	2.47	100.74
Summer (Boro) Rice	15.57	18.5	2.80	43.58
Total Rice	59.36	70.7	2.53	150.37
Wheat	33.07	39.4	2.49	7.65
Total Cereals	63.53	75.7	2.55	161.67
Total Pulses	1.84	2.2	0.71	1.30
Total Foodgrains	65.37	77.9	2.49	162.97
Total Oilseeds	7.04	8.4	0.83	5.83
Potato	3.86	4.6	10.68	41.21
Jute	5.84	7.0	0.01	78.73
Total (Foodgrains, oilseed, potato, jute)	83.95	100.0		

Source: Government of West Bengal, *Economic Review*.

The district-wise cropping intensity with gross cropped area and net sown area is presented in the following Table 1.2. The cropping intensity is lowest in the district of Purulia. As Table reveals the cropping intensity is less compared to state average in the districts in the dry zones like Bankura, Purulia and Paschim Medinipur. The major way to improve this is to provide more irrigation. As there is less possibility of well irrigation in these regions, irrigation through rainwater harvesting technique should be promoted in a bigger way.

Table 1.2: Gross Cropped Area, Net Sown Area and Cropping Intensity in the Districts of West Bengal during 2008-09

(Area in '000 hectares)

District	Cultivable Area	Gross Cropped Area	Net Sown Area	Cropping Intensity %
Darjeeling	158	194	132	146
Jalpaiguri	353	551	336	164
Coochbehar	258	552	252	219
Uttar Dinajpur	279	508	277	184
Dakshin Dinajpur	188	327	186	176
Malda	281	451	212	212
Murshidabad	401	938	400	235
Nadia	299	697	289	241
North 24 Parganas	264	511	257	198
South 24 Parganas	382	540	374	144
Howrah	85	158	80	199
Hooghly	219	555	215	258
Burdwan	468	862	452	190
Birbhum	336	568	320	177
Bankura	390	514	347	148
Purulia	439	374	317	118
Paschim Medinipur	595	982	560	175
Purba Medinipur	294	520	288	180
West Bengal	5689	9802	5294	185

Source: Government of West Bengal, *Economic Review*.

1.1.3 Types of Irrigation

Irrigation is an artificial means of watering the crops or plants or an art of supplying water to the crops (*Gurjar, 2008*). Irrigation in India is divided into four categories as – Canal irrigation, River-lift irrigation (RLI), Tube well irrigation and Tank irrigation.

- *Canal Irrigation*: Surface water stored in reservoirs or rivers and streams are distributed for irrigation through the canal irrigation network.
- *River-Lift Irrigation (RLI)*: Water lifted directly from rivers and streams from suitable points at different segments of canal system. The system has usually designed to serve a group of farmers and the project is owned and operated by government agencies or by farmers' co-operatives.
- *Irrigation Well*: Well consists of dug wells, shallow tube wells and deep tube wells. Diesel engines and electric motors almost exclusively do water lifting from shallow tube wells and deep tube wells.

- *Tank/Reservoir*: An artificially impounded body of water constructed for the purpose of storage, regulation and control of water. Tanks constructed exclusively for groundwater recharge are called percolation tanks.

1.1.4 Factors Affecting Irrigation Development

Different systems of irrigation are found in different geographical contexts. Factors affecting irrigation development are as follows:

1. *Topography and Slope of Soil*: The type of irrigation depends very much on the topography or structure of the ground level. Tanks are more suitable in stony areas where tube well irrigation is costly. Ground level plays an important role in construction of wells. If ground level is hard and rocky, deeper wells cannot be constructed. Similarly, slope of ground level also affects irrigation development along with agriculture. Canal system has developed according to the slope.
2. *Structure, Depth and Permeability of Soil*: Water can be absorbed in the soil according to the structure of the soil. If there is lesser depth, it affects permeability and it can create a condition of water logging.
3. *Depth of Groundwater*: Quantity of groundwater and its depth also affects irrigation development.
4. *Climate*: Quantity and duration of rainfall affect irrigation development. On this basis, rainwater flow and seepage is affected. Evaporation and transpiration occurs according to temperature. Speed of air affects evaporation. Thus, climate has an important effect on irrigation development.
5. *Social and Economic Factors*: Social reasons also affect irrigation development. Scattered fields or their consolidations directly affect the irrigation development. Problems of resettlement also affect some big or small irrigation development projects.

1.2 Irrigation Development in West Bengal

1.2.1 Irrigation Development in West Bengal Compared to Other States

In the post-independence period irrigation works were done based on three classifications: (i) major irrigation: culturable command area of

10,000 he or more, (ii) medium irrigation projects: culturable command area between 2000 and 10,000 ha, and (iii) Minor irrigation projects: below 2,000 ha. Minor irrigation has further classified into surface water minor irrigation and ground water minor irrigation schemes. As water is the responsibility of the state in the Indian Constitution, states are primarily responsible for planning, implementation and funding management. The three main institutions involved in irrigation development are Ministry of Water Resources, the Planning Commission and the Ministry of Agriculture.

The development of irrigation is not same across the states. There is also the regional disparity of irrigated area across states. Percentage share of gross irrigated area in different states in India had shown in Table 1.3. West Bengal has 9.4 per cent of the total irrigated area of the country in 2002-03. As Table 1.1 shows the percentage share of irrigation has increased from 4.6 per cent in 1998-99 to 9.4 per cent in 2002-03 which shows improvement in irrigated area in the state. Uttar Pradesh has the highest share of irrigated land followed by Punjab.

Per capita state-wise land use statistics have been given in Table 1.4. Per capita net irrigated area is 35.77 hec per 1000 population in West Bengal which is less than the country average of 51.06 hec. The net irrigated area as a percentage of net sown area is 54 per cent and West Bengal's position is 7th out of 34 states in India. In terms of net sown area. West Bengal's position is 27th which indicates the importance of irrigation development in the state.

1.2.2 Irrigation Demand in West Bengal

Water is required for different sectors like agriculture, industry, energy etc. Water requirement for agriculture is high compared to other sectors. The total water requirement in West Bengal is projected to increase from 10.85 mham to 17.95 mham the requirement of water for agriculture is projected to increase in the year 2025 to 62.4 per cent from 49.6 per cent in 2000 (*WBPCB*).

Table 1.6 shows the water requirement and deficit in West Bengal (Mham). From Table 1.6, we see that water deficit also will be more in the year 2025 than the year 2000.

As Table 1.5 shows the irrigation demand is projected to be very high for agriculture. Water is needed in agriculture mainly to meet the demands of evaporation (E), transpiration (T) and metabolic needs

Table 1.3: Percentage Share of Gross Irrigated Area in Different States in India

States	*1998-99*	*1999-00*	*2000-01*	*2001-02*	*2002-03*
Andhra Pradesh	11.2	10.5	11.1	9.8	8.6
Arunachal Pradesh	0.1	0.1	0.1	0.1	0.1
Assam	1.1	1.0	0.4	0.4	0.4
Bihar	0.0	0.0	0.0	8.0	8.7
Bihar (Erstwhile-till 2001)	8.8	8.8	9.0	0.0	0.0
Chhattisgarh	2.3	2.3	1.9	2.2	2.2
Delhi	0.1	0.1	0.1	0.1	0.1
Goa	0.1	0.1	0.1	0.1	0.1
Gujarat	7.1	6.6	6.8	6.3	5.9
Haryana	9.3	9.4	9.8	9.4	9.9
Himachal Pradesh	0.3	0.3	0.3	0.3	0.3
Jammu & Kashmir	0.8	0.8	0.8	0.8	0.8
Karnataka	5.8	5.8	6.1	5.5	5.4
Kerala	0.8	0.9	0.9	0.8	0.8
Madhya Pradesh	10.4	10.6	8.0	8.6	8.8
Maharashtra	6.9	6.9	7.1	7.0	7.7
Manipur	0.1	0.1	0.1	0.1	0.1
Meghalaya	0.1	0.1	0.1	0.1	0.1
Mizoram	0.0	0.0	0.0	0.0	0.0
Nagaland	0.1	0.1	0.1	0.1	0.2
Orissa	4.3	4.6	4.0	4.5	3.2
Punjab	13.8	13.7	15.0	13.5	14.4
Rajasthan	12.5	12.7	11.5	11.9	10.0
Sikkim	0.0	0.0	0.0	0.0	0.0
Tamil Nadu	6.7	6.5	6.5	6.0	5.0
Tripura	0.1	0.1	0.1	0.1	0.1
Uttar Pradesh	32.6	32.3	34.0	32.0	33.8
Uttarakhand	1.0	1.0	0.0	1.0	1.1
West Bengal	4.6	4.5	6.3	8.7	9.4
Total	100.0	100.0	100.0	100.0	100.0

Source: Central Water Commission.

of the plants, altogether known as consumptive use (CU) for plants. From supply side, this water requirement (WR) can be written by the following equation:

WR = Irrigation requirement (IR) + Effective rainfall (ER) + Ground water (S)

Table 1.4: State-wise Per Capita Land Use Statistics (hectare/thousand person), 2004

State/UT	Net Sown Area (NSA)		Net Irrigated Area (NIA)		NIA as a % of NSA	
	Hector per Thousand Persons	Rank	Per Thousand Persons	Rank	Percentage	Rank
Andaman	43.48	30	0	34	0.0	35
Andhra Pradesh	128.12	13	46.02	8	35.9	13
Arunachal	143.73	12	36.81	12	25.6	19
Assam	99.5	16	6.17	30	6.2	33
Bihar	65.1	28	39.12	10	60.1	6
Chandigarh	2	34	1	33	50.0	8
Chhittisgarh	218.18	3	49.76	7	22.8	22
Dadra	95.83	18	29.17	17	30.4	16
Daman	10.31	33	0	35	0.0	34
Delhi	1.78	35	1.65	32	92.7	2
Goa	100.07	15	17.03	24	17.0	27
Gujarat	180.56	5	57.16	6	31.7	15
Haryana	157.42	9	132.25	2	84.0	3
Himachal Pradesh	86.38	22	19.65	21	22.7	23
Jammu & Kashmir	70.33	25	28.9	19	41.1	11
Jharkhand	62.32	29	5.78	31	9.3	31
Karnataka	179.28	6	43.4	9	24.2	20
Kerala	66.96	26	11.74	29	17.5	26
Lakshadweep	43.48	31	14.49	27	33.3	14
Madhya Pradesh	233.49	2	87.98	3	37.7	12
Maharashtra	171.53	7	28.97	18	16.9	28
Manipur	96.4	17	17.77	22	18.4	25
Meghalaya	94.23	20	24.91	20	26.4	18
Mizoram	106.18	14	17.33	23	16.3	29
Nagaland	147.49	11	32.4	16	22.0	24
Orissa	152.19	10	34.82	14	22.9	21
Pondicherry	20.61	32	16.68	25	80.9	4
Punjab	167.15	8	159.39	1	95.4	1
Rajasthan	289.98	1	87.34	4	30.1	17
Sikkim	195.73	4	16.01	26	8.2	32
Tamil Nadu	73.16	24	33.51	15	45.8	9
Tripura	84.24	23	12.03	28	14.3	30
Uttar Pradesh	95.32	19	70.25	5	73.7	5
Uttarakhand	86.93	21	38.76	11	44.6	10
West Bengal	66.28	27	35.77	13	54.0	7
India	130.55		51.06		39.1	

Source: Central Water Commission.

Table 1.5: Percentage of Projected Water Requirement in West Bengal

Sectors	2000	2011	2025
Agriculture	49.6	59.3	62.4
Domestic	2.4	2.2	2.2
Industry	2.4	2.9	3.4
Power (Thermal)	2.9	—	—
Inland Navigation	33.5	27.9	20.6
Forestry	0.1	0.1	5.7
Ecology, Environment and Others	9.2	7.7	5.7
Total (mham)	100.0	100.0	100.0

Source: WBPCB (2009).

Table 1.6: Water Requirement and Deficit in West Bengal (Mham)

Sl. No.	Year	Water Requirement	Deficit	% deficit
1.	2000	10.85	4.08	37.60
2.	2011	13.02	6.25	48.00
3.	2025	16.60	9.83	59.22

Source: WBPCB (2009).

1.2.3 District-wise Irrigation Development in West Bengal

Irrigation development in West Bengal is also biased in favour of well irrigation. Out of 30.48 lakh hectares of irrigated land in West Bengal in 2005-06, source-wise irrigated land are as follows: Government canal = 10.65 lakh hectares, Tube well = 11.53 lakh hectares and tanks = 2.45 lakh hectares. Tank irrigation plays an important role in the districts like Bankura, Purulia, Dakshin Dinajpur etc. (Table 1.7).

District-wise percentage share of different sources of irrigation in West Bengal is given in Table 1.8. The shares for different sources of irrigation for West Bengal are as follows: Government canal -- 35 per cent, Tank – 8 per cent, Tube Well – 37.8 per cent, River Lifting Irrigation (RLI) – 5.7 per cent and others – 13.5 per cent. The shares of tank irrigation are comparatively high in Purulia, Dakshin Dinajpur, Bankura etc.

District-wise cultivated area covered by irrigation in West Bengal has been shown in Table 1.9. The percentage of cultivable area covered by minor irrigation in West Bengal is 23.5 per cent and the percentage of total irrigated area is 35.6 per cent. The percentage of irrigated

Table 1.7: District-wise Irrigated Area in West Bengal (2005-06)

(in thousand hectares)

Districts	*Govt. Canal*	*Tank*	*Tube Well*	*RLI*	*Others*	*Total*
Hooghly	99.8	37.5	155.0	42.9	0.0	335.1
Jalpaiguri	58.4	2.1	6.7	8.9	17.6	93.6
Darjeeling	2.4	0.0	2.0	4.2	0.3	8.9
Birbhum	159.9	25.3	49.4	2.1	55.2	291.8
Uttar Dinajpur	2.6	6.0	132.2	6.9	0.0	147.7
24 Parganas North	7.5	15.1	117.4	11.9	3.7	155.7
Midnapore East	63.6	26.3	75.8	3.7	39.8	209.1
Howrah	35.5	9.0	4.8	2.3	0.0	51.6
Malda	0.0	1.3	90.1	9.1	24.9	125.4
Coochbehar	1.7	5.9	72.5	14.3	12.1	106.5
Nadia	0.0	0.0	192.2	11.1	9.7	213.0
Mursidabad	32.9	7.8	11.4	11.1	141.4	204.7
Burdwan	283.7	0.0	22.4	11.9	0.0	318.0
Dakshin Dinajpur	0.0	10.4	12.1	5.8	0.0	28.3
Purulia	30.3	28.3	0.0	1.3	12.2	72.1
24 Parganas South	44.9	12.3	14.0	4.1	34.3	109.5
Bankura	176.3	33.1	50.1	5.4	8.7	273.6
Paschim Medinipur	66.1	24.7	145.2	15.7	52.3	304.0
West Bengal	1065.5	245.0	1153.2	172.6	412.1	3048.5

Table 1.8: District-wise Percentage of Irrigation Development in West Bengal (2005-06)

Distrcts	*Govt. Canal*	*Tank*	*Tube Well*	*RLI*	*Others*	*Total*
Hooghly	29.8	11.2	46.2	12.8	0.0	100.0
Jalpaiguri	62.4	2.3	7.1	9.5	18.8	100.0
Darjeeling	27.2	0.0	22.5	47.1	3.2	100.0
Birbhum	54.8	8.7	16.9	0.7	18.9	100.0
Uttar Dinajpur	1.8	4.1	89.5	4.7	0.0	100.0
24 Parganas North	4.8	9.7	75.4	7.7	2.4	100.0
Midnapore East	30.4	12.6	36.2	1.7	19.0	100.0
Howrah	68.8	17.4	9.3	4.5	0.0	100.0
Malda	0.0	1.0	71.8	7.3	19.9	100.0
Coochbehar	1.6	5.5	68.1	13.4	11.3	100.0
Nadia	0.0	0.0	90.2	5.2	4.6	100.0
Mursidabad	16.1	3.8	5.6	5.4	69.1	100.0
Burdwan	89.2	0.0	7.1	3.7	0.0	100.0
Dakshin Dinajpur	0.0	36.7	42.7	20.7	0.0	100.0
Purulia	42.0	39.2	0.0	1.8	16.9	100.0
24 Parganas South	41.0	11.3	12.8	3.7	31.3	100.0
Bankura	64.4	12.1	18.3	2.0	3.2	100.0
Paschim Medinipur	21.7	8.1	47.8	5.2	17.2	100.0
West Bengal	35.0	8.0	37.8	5.7	13.5	100.0

Source: Government of West Bengal, District Statistical Handbooks.

Table 1.9: Water for Agriculture in the Districts of West Bengal

District	Total area (sq. km.)	Cultivable area (sq. km.)	Percentage of cultivable area covered by minor irrigation (2000-01)			Percentage of cultivable area covered by major non-monsoon irrigation (2004-05)	Sum of Percentages of cultivable area covered by minor and major non-monsoon irrigation
			Ground water	Surface water	Total		
Bankura	6882	3958	12.9	25.8	38.7	7.2	45.9
Burdwan	7024	4915	34.3	7.4	41.7	12.6	54.2
Birbhum	4545	3441	14.6	16.0	30.6	5.1	35.6
Coochbehar	3387	2694	23.4	3.3	26.6	0	26.6
Darjeeling	3149	1622	5.0	9.0	14.0	0	14.0
Daskhin Dinajpur	2219	1952	25.4	11.5	36.9	0	36.9
Hooghly	3149	2302	40.7	12.9	53.6	6.5	60.1
Howrah	1467	941	8.7	62.0	70.6	0.3	70.9
Jalpaiguri	6227	3648	12.5	8.8	21.4	0	21.4
Maldah	3733	2853	34.2	8.5	42.7	0	42.7
Murshidabad	5324	4089	41.7	6.1	47.8	1.0	48.8
North 24 Parganas	4094	2686	38.4	3.0	41.4	0	41.4
Nadia	3927	3145	38.1	6.0	44.1	0	44.1
Paschim Medinipur	9286	6104	23.2	9.8	33.0	4.0	37.0
Purba Medinipur	4795	3023	23.2	9.8	33.0	0	33.0
Purulia	6259	4614	1.1	22.2	23.3	1.7	25.0
24 South Parganas	9960	4005	4.0	19.6	23.5	0	23.5
Uttar Dinajpur	3140	2811	42.6	3.2	45.8	0	45.8
West Bengal	88752	58802	23.5	12.2	35.6	2.7	38.3

Source: WBPCB 2009.

area in the districts of Bankura, Purulia and Paschim Medinipur are respectively 38.7 per cent, 23.3 per cent and 33 per cent of the cultivated areas and most of the irrigation in Bankura and Purulia is surface irrigation type.

Utilisation of irrigation potential is an important factor for agricultural development. Creation and utilization of irrigation in West Bengal have presented in Table 1.10. It is seen from Table 1.10 that percentage of utilization over creation of major and medium irrigation was 74.98 in the year 1996-97, which increased in the year 2007-08 to 81.73, and in case of minor irrigation it was 82.81 per cent in the year 1996-97, which decreased in the year 2007-08 to 81.64 per cent in West Bengal.

Table 1.10: Creation and Utilization of Irrigation in West Bengal

Year	*Potential created upto the year*			*Percentage of utilization Over creation*		
	Major & Medium ('000 Hec)	*Minor ('000 Hec)*	*Total ('000 Hec)*	*Major & Medium*	*Minor*	*Total*
1996-97	1335.5	3089.0	4424.5	74.9	82.8	80.4
1997-98	1356.5	3164.0	4520.5	89.6	82.7	84.8
1999-00	1429.7	3364.0	4793.7	84.3	82.7	83.2
2000-01	1477.9	3427.1	4905.1	70.3	83.6	79.6
2001-02	1532.9	3564.0	5096.9	67.4	82.8	78.1
2002-03	1549.4	3629.0	5178.4	69.4	82.8	78.8
2003-04	1552.2	3699.0	5251.2	69.1	82.7	78.7
2004-05	1554.7	3764.0	5318.7	75.6	82.4	80.4
2005-06	1560.7	3814.0	5374.7	74.3	82.1	79.9
2006-07	1566.1	3864.0	5430.1	77.1	81.9	80.5
2007-08	1570.8	3930.2	5501.1	81.7	81.6	81.6

Source: WBPCB (2009).

1.2.4 Irrigation Expenditure in West Bengal

Irrigation accounts for a significant amount of state budget in West Bengal. In Table 1.11, we have presented the expenditure on irrigation in West Bengal. Total expenditure on irrigation can be decomposed into major and medium irrigation, minor irrigation, Command area development, flood control. Minor irrigation expenditure has increased from Rs. 36.52 crore in 1985-86 to Rs. 239.11 crore in 2004-05. Major and medium irrigation mainly comprises canal irrigation. Minor irrigation comprises well irrigation and tank irrigation. There is no separate data available for tank irrigation. As well irrigation is mainly private in nature, minor irrigation expenditure may contain a high share of tank irrigation expenditure.

We have also calculated the percentage share of minor irrigation expenditure over the same period shown in Table 1.12. The percentage share of minor irrigation expenditure is seen to have improved over the years. It is to be mentioned that though the share of major irrigation is still very high, the area under major and medium irrigation is not improving.

We have also calculated the minor irrigation expenditure as a percentage of SDP and as a percentage of Budgeted expenditure over the periods 1994-95 to 2004-05 shown in Table 1.13. As has been revealed from table that expenditure on minor irrigation has been reduced from 0.18 per cent of SDP in 1994-95 to 0.126 per cent of SDP in 2004-05. This indicates that there is less emphasis on the maintenance of tank irrigation. The tank irrigation may have falling tendency because of this. It may also be observed that total irrigation expenditure as a percentage of SDP and as a percentage of budgeted expenditure has a declining tendency.

The line diagrams showing minor irrigation expenditure as a percentage of SDP and budgeted expenditure have been presented in the Fig. 1.1.

1.3 Concept and Practice of Tank Irrigation

Tanks in Indian context are inextricably linked to the socio-cultural aspects of rural life and have historically been an independent part of village habitat sustaining its socio-ecological balance (*Reddy 2008*). Tanks were mostly constructed in South and Eastern India in the ancient time mainly for multiple uses like irrigation, livestock and human uses. The predominance of tanks in the Deccan Plateau is its tracts with undulating topography and rocky substrata which is ideally suited for locating tank in the valley depression and carrying out gravity irrigation. In case of the plains of south Bihar and Chhota Nagpur plateau, the indigenous system of the *ahar-pyne* is the outcome of the natural condition and physical configuration of the area. Similarly, somewhat steeper gradients of West Bengal offer greater scope for gravity flow irrigation from smallscale reservoirs (tanks) (*ADB 2007*). Tank irrigation has a rich heritage on account of long historical antecedents in eastern India, consisting of eastern Uttar Pradesh (UP),

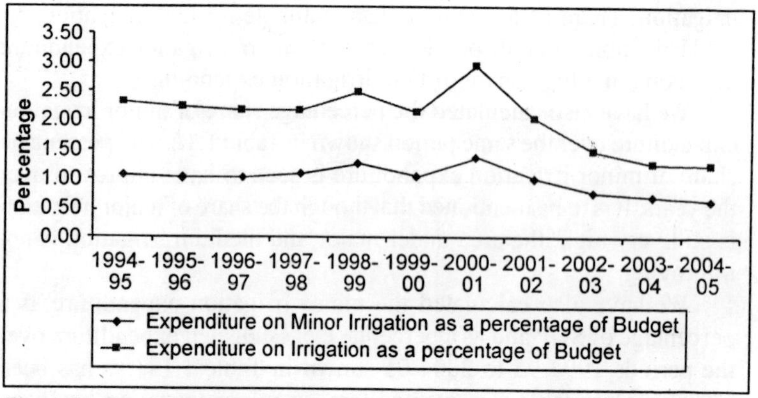

Fig. 1.1: Percentage Expenditure on Minor Irrigation and Total Irrigation in West Bengal

Table 1.11: Expenditure on Irrigation in West Bengal

(Rs. Lakh)

Years	*Major and Medium Irrigation*	*Minor Irrigation*	*Command Area Development*	*Flood Control*	*Total*
1985-86	2894	3653	71	1835	8452
1986-87	3534	3666	86	2167	9453
1987-88	3622	4542	89	2274	10527
1988-89	4127	5128	102	2561	11918
1989-90	4659	5861	123	2940	13584
1990-91	5698	7857	134	2919	16608
1991-92	5815	7266	108	2663	15851
1992-93	6096	8142	117	2714	17069
1993-94	7024	10218	138	3732	21112
1994-95	7653	9940	138	3903	21634
1995-96	8562	11223	166	4475	24427
1996-97	9979	13016	196	4596	27788
1997-98	11133	15970	215	4980	32299
1998-99	14077	20748	341	5876	41043
1999-00	17189	25855	345	10231	53621
2000-01	19087	33195	384	19164	71830
2001-02	16139	28797	384	14178	59498
2002-03	15182	23238	382	6640	45442
2003-04	14218	24101	374	5682	44375
2004-05	16924	23912	375	6564	47774

Source: Government of West Bengal, Statistical Abstract, Various Issues.

Table 1.12: Percentage of Expenditure on Irrigation in West Bengal

(Percentage)

Years	*Major and Medium Irrigation*	*Minor Irrigation*	*Command Area Development*	*Flood Control*	*Total*
1985-86	34.2	43.2	0.8	21.7	100.0
1986-87	37.4	38.8	0.9	22.9	100.0
1987-88	34.4	43.1	0.8	21.6	100.0
1988-89	34.6	43.0	0.9	21.5	100.0
1989-90	34.3	43.1	0.9	21.6	100.0
1990-91	34.3	47.3	0.8	17.6	100.0
1991-92	36.7	45.8	0.7	16.8	100.0
1992-93	35.7	47.7	0.7	15.9	100.0
1993-94	33.3	48.4	0.7	17.7	100.0
1994-95	35.4	45.9	0.6	18.0	100.0
1995-96	35.1	45.9	0.7	18.3	100.0
1996-97	35.9	46.8	0.7	16.5	100.0
1997-98	34.5	49.4	0.7	15.4	100.0
1998-99	34.3	50.6	0.8	14.3	100.0
1999-00	32.1	48.2	0.6	19.1	100.0
2000-01	26.6	46.2	0.5	26.7	100.0
2001-02	27.1	48.4	0.6	23.8	100.0
2002-03	33.4	51.1	0.8	14.6	100.0
2003-04	32.0	54.3	0.8	12.8	100.0
2004-05	35.4	50.1	0.8	13.7	100.0

Source: Own Estimation based on Government of West Bengal, Statistical Abstract, Various Issues.

Table 1.13: Expenditure on Minor Irrigation and Total Irrigation (% in SDP and Budgeted Expenditure) in West Bengal

Years	*Expenditure on Minor Irrigation as a percentage of SDP*	*Expenditure on Irrigation and Flood Control as a percentage of SDP*
1994-95	0.18	0.40
1995-96	0.17	0.36
1996-97	0.17	0.37
1997-98	0.18	0.36
1998-99	0.20	0.39
1999-00	0.22	0.46
2000-01	0.26	0.56
2001-02	0.20	0.42
2002-03	0.15	0.30
2003-04	0.14	0.26
2004-05	0.13	0.25

Source: Own Estimation based on Government of West Bengal, Statistical Abstract, Various Issues.

south Bihar (plains as well as plateau), West Bengal and Orissa. In the eastern region, tanks and ponds have been an important supplementary source of irrigation over centuries. South India has more tanks because of its geography, climate, and terrain situations. The geological formation of south India is of hard granite gneisses, which helps reduce deep percolation from tanks and ponds. This may be yet another reason for the existence of more tanks in southern peninsular India than in the north. The north-east monsoon is more active in the coastal districts of Tamil Nadu and Andhra Pradesh and they also have the maximum number of tanks. The southern states of Andhra Pradesh, Karnataka, and Tamil Nadu put together have around 143,000 tanks, constituting nearly 50 per cent of the tanks in India (*ADB 2007*).

India has an age-old tradition of rain-water harvesting. The tradition developed over centuries in ecologically sound way where rural and urban communities played an active role in water management. When the British took control of India, the Government replaced traditional decentralized system with centralized ones with Government taking the main role of provider. It continued even after the British left which is marked as the end of a great tradition (*CSE 2001*). In recent years some successful efforts are going on to rejuvenate water harvesting structures. Some successful such programmes are new *ahar* and *pyne* system in Bihar, Jhabua in Madhya Pradesh.

In West Bengal tank irrigation was most widely practised indigenous irrigation method since independence. It was mostly seen in the dry western plateau districts of Purulia and Bankura. Depending on the slopes of terrain crescent shaped tanks with earthen embankments were constructed at different levels. This practice of harnessing run off and rain water is known as bundh irrigation. Trees were observed to be planted along such reservoirs to prevent soil erosion and reducing evaporation. In many cases these expensive bundhs were created by local rulers. Another type of irrigation often seen in these regions is *jhor* bund-irrigation, where earthen bunds are constructed across small seasonal streams (*jhors*) along different reaches. In times of scarcity, the water of the upper reservoir is first utilized. Water is distributed from these *jhors* to fields through gravity flow channels. Rectangular tanks are mostly seen in relatively flat terrain. In saline water regions as in the district of 24 Parganas, elongated tanks called *khals* are observed which tap mainly rainwater and floodwater. As these *khals* are not connected to river streams, these are termed as tanks (*Chatterjee 1995*).

1.3.1 Advantages of Irrigation by Tanks

About 53 per cent of geographical area falls under semi-arid region where communites have practised water conservation through tanks. Tank irrigation, in certain parts of India, provides a good alternative for irrigation development. Tanks can have a wider geological distribution than large-scale projects. Income distribution and employment generation effects are not limited to one area. Tank investments tend to be less capital intensive, have fewer negative environmental impacts and can better involve local communities in improvement and construction works. Currently the tank irrigation potential is underutilized due to problems relating to rainfall, poor tank structures, and lack of collective tank management. The following are the advantages of tank irrigation (*Gulati, et al.* 1994, *Pain et al.* 2008, *Palanasai* 2000, *Reddy* 2009):

1. Tanks have wider geographical distribution than the large-scale projects.
2. It is a cheap and popular method of irrigation. It is cheaper than other sources because it involves less capital. The costs of new medium sized surface irrigation projects are about Rs. 53,000 per hectare.
3. It has fewer negative environmental impacts and it maintains/ improves the level of groundwater through recharging of groundwater.
4. Where construction of wells is very costly due to hard rock surface, irrigation by tank is cheaper. In some cases, construction of wells is not possible at all.
5. Expenditure is much less as compared to major/medium irrigation projects.
6. Use of additional water from rainfall is possible through tanks.
7. Development of tank irrigation needs shorter gestation period. A tank can be constructed within a year.
8. It involves local community and the benefits are accrued to the community.
9. It does not require big amount of land and so the problem of displacement is not present here.
10. The development of tank irrigation may generate benefits to the target group. But large projects benefit only one section of a district or state and are many times limited by physical characteristics, i.e. there are only a limited number of large dam sites.

11. Tank irrigation is a cheap mode of irrigation. The rapid increase in electricity price and the diesel price has slowed down the process of ground water irrigation in recent years. The power shortages have also constrained groundwater development. In addition, ground water irrigation is limited to the availability of ground water. In view of these problems, it is felt that tank modernization could become a viable source for meeting future demands for irrigated acreage.
12. The threat of climate change is now almost an established fact. It has been predicted that global warming will lead to intense rainfall in fewer spells with floods and droughts will become more intense (*GoI 2008*). According to IPCC, rainfed agriculture in dry tropics will be the most vulnerable. An increase in mean temperature will lead to more evapo-transpiration leading to more crop water requirement. Increasing soil moisture through tanks will be more needed.

In spite of the above advantages of tank irrigation, development of tank irrigation deteriorated in almost every state in India. The possible causes behind this are discussed below:

I.3.2 Degradation of Traditional Methods

The main sources of irrigation in India are — canals, tanks and wells including tube wells. A substantial progress has been made during the planned period in the irrigation development in the country. The gross irrigated area in India was 87.26 million hectares in 2007-08 (*CWC 2010*) which has been increased from 31.103 million hectares in 1970-71. But the importance of tank irrigation has decreased over the years. The share of tank irrigated area has decreased from 13.2 per cent in 1970-71 to 9.6 per cent in 2007-08. Irrigation by tanks is highest in the state of Tamil Nadu at 17.7 per cent followed by Jharkhand at 16.2 per cent (*CWC 2010*). Of the three main sources of irrigation, the tube well irrigation has seen a tremendous growth in the last few decades. The share of well irrigated area has increased from 38.2 per cent in 1970-71 to 60.7 per cent in 2007-08. The growth of tank-irrigated area in India was positive for the period 1950-51 to 1960-61, after that the growth of tank-irrigated area declined consistently for three consecutive decades and growth rate was positive in the decade of 1990s (*Narayanmoorthy and Deshpande 2003*). The percentage share of tank irrigated area over the period of 1970-71 to 2007-08 is presented in Fig. 1.2.

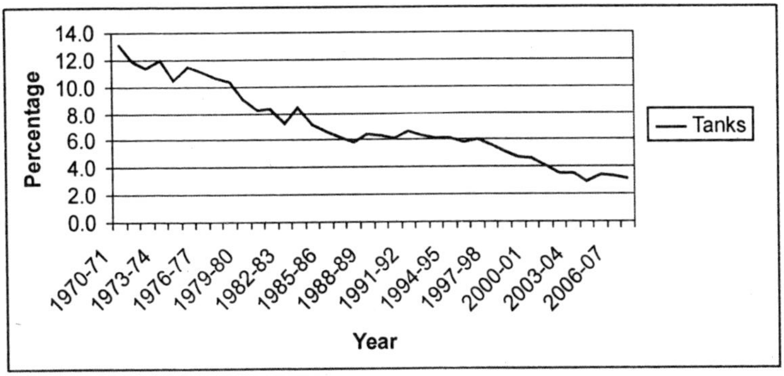

Source: Central Water Commission, Water and Related Statistics, Various Years.

Fig. 1.2: Year-wise Percentages Share of Irrigated Area by Tanks in India

The main constraint operating against tank irrigation development is inadequate water supply during winter and summer seasons. The causes of degradation of tanks as found in literature are as follows (*Narayanmoorthy and Deshpande 2005*, *ADB 2007*, *Shah 2003*, *Vaidyanathan 2006*, *Agarwal 2001*).

- The massive increase in demographic pressure on land has contributed significantly to extension of cultivation to marginal lands, encroachment of common lands, government and forestlands, and deforestation. One of the most important reasons for the decline in tank irrigation is the disappearance of village institutions that were managing the tanks. The village level organizations have undergone changes because of a vast increase in the number of irrigators, shift in land control from a small to a large number of landowners, changes in the attitude of farmers, and the spread of well irrigation in the tank command. Today, tanks are viewed more as an "open access resource"; tank communities have lost their capacity and the will to mobilize resource and labour to undertake regular maintenance activity to enforce norms and obligations. New technologies have become available which made it possible to augment water supply beyond the limits of existing traditional devices. The introduction of drilling methods and energized pumping permit tapping deeper strata of ground water, which is impossible with traditional devices. This in the case of irrigation, tube wells and electrical/diesel pumps enable larger volumes of water to be lifted per unit of time.

- Costs of ground water have invariably been underestimated and environmental costs altogether ignored even as benefits have been consistently overstated. There are far many instances of bad management and indiscriminate irrigation leading to land degradation. Policies (like cheap electricity and diesel, low interest credit and lax loan recovery) have encouraged inefficient use of water and overexploitation of ground water in most parts of the country.
- The construction of dams/reservoirs in the upper catchment results in a less quantity of water reaching the downstream tanks.
- The faulty design features of the newly constructed tanks resulted in the low performance of the tanks.
- Due to government emphasis on more convenient systems of centralized storage arrangements in the form of water reservoirs and canals, extension of traditional to individual beneficiary schemes by the government simultaneously lead to neglect of active community participation in maintenance of traditional systems.
- Emergence of livelihood patterns and urbanisation that are increasingly independent of tank based occupation.
- Inadequate government expenditure for the construction/ renovation of tanks.
- Declining attention and investment on tanks as there is little scope for rent.

1.4 Tank Irrigation in West Bengal

1.4.1 Planned Tank Irrigation Development in West Bengal

The importance of tank irrigation development was felt much earlier. Tank Irrigation comes under Minor irrigation. Minor irrigation facilities in West Bengal are created mainly by Water Investigation and Development Department (WIDD), Panchayat and Rural Development Department that mainly operate the District level bodies, DRDA organizations, through grants-in-aid programme, Jawahar Gram Samridhi Yojana (SGSY), Sampoorna Grameen Rojgar Yojana, Rashtriya Sama Vikash Yojana, different Zilla Parishad programmes, Department of Agriculture, other Backward Class Welfare Department.

Water Investigation and Development Department (WIDD) under Government of West Bengal came into being in the present form and style from the year 1995. Before the early eighties, minor irrigation structures used to be constructed by the Engineering Wing attached to the Agriculture Department. With the increasing need for installation of more and more minor irrigation structures in pursuance of the "Grow more food" campaign, a separate department for Minor Irrigation Wing was created with a Special Secretary as its Head, which was renamed in 1995 as "Water Investigation and Development Department", although the objective of the present Department has not been changed at all at any stage. The Department is also engaged in implementation of various minor irrigation schemes under centrally sponsored programme. Principal Secretary mostly does interactions with the officials of Government of India under concerned ministries like Ministry of Water Resources, Planning Commission, Central Water Commission, and NABARD etc. Presently, this Department is the controlling authority in respect of two major directorates; two State owned Corporations and four CAD Authorities as noted hereunder:

1. Water Resources Development Directorate (WRDD).
2. Command Area Development Authorities (CADAs), viz.
   (i) Teesta Command Area Development Authority,
   (ii) Damodar Valley Command Area Development Authority,
   (iii) Kangsabati Command Area Development Authority and
   (iv) Mayurakshi Command Area Development Authority.
3. State Water Investigation Directorate (SWID).
4. West Bengal State Minor Irrigation Corporation Ltd.
5. West Bengal Agro-industries Corporation Ltd.

WRDD is mainly engaged in developing existing potential of water resources available in the state for irrigation purposes and construction and maintenance of various minor irrigation structures for utilization of ground water as well as surface water.

Meteorologically West Bengal gets abundant rainfall, yet the state is unable to make full use of it due to some natural constraints. Rapid rise in demand of water in all sectors of domestic, industrial and agricultural uses during the last decade has started throwing up several problems in water resources management. So the optimal management of available Water Resources has become a major issues world over. Due to occurrence of recurrent droughts, advent of high yielding varieties and introduction of an incentive oriented agricultural pricing policy paved the way for irrigated agriculture. The spatial and temporal variability of rainfall along with high evapotranspiration and runoff is posing a major challenge to the scientific community. As a result, development of agriculture has been essentially dependent on irrigation, not only in Rabi / Boro season but also in Kharif season. In the event of drought or inadequate rainfall, harnessing of Water resources to meet various sectoral requirements has been our prime concern, which is the foundation of our agricultural *vis-à-vis* economic growth and welfare of state. Water Resources are being utilized in the State from medium and major sized irrigation schemes whose culturable command area above 2000 he. and from minor irrigation schemes with the C.C.A. up to 2000 he. as per the criteria laid down by the Planning Commission, Government of India.

Next to land reforms, as presented in various issues of Economic Review, the Government of West Bengal lays importance to the development of irrigation potential in the State mainly through the creation of minor irrigation potential. According to Planning Commission norms, irrigation schemes with cultural command area of 2 thousand hectares or less are considered as minor irrigation schemes. The State Government lays emphasis on the spread of minor irrigation in the State, not only because it is, in many cases, local resource and labour intensive, but because the utilization of minor irrigation potential created is much higher than that from medium and large irrigation projects. Moreover, it is easier to ensure that the beneficiaries of a minor irrigation source are from the target group than in the case of larger irrigation systems.

A new concept of management of minor irrigation facilities had evolved in the State by handing over the management of low capacity deep tube-wells and shallow tube-wells to the beneficiaries through *Panchayat Samities*.

A large number of minor irrigation scheme are being executed by the different Zilla Parisads under RIDF and Grants-in-Aid Programme. Different schemes on Survey and Investigation of Ground Water and Surface Water Resources of Development minor irrigation in the State are continuing.

In 59 blocks in five laterite districts—Purulia, Bankura, Birbhum, Midnapore (West) and part of Burdwan district—efficient and economical utilization of land and water resources are being attempted through popularization of scientific dry land farming in about 301 selected micro watersheds and also outside these watersheds. As mono-crops is the usual system and the productivity of different crops is low in these areas, farmers are encouraged to adopt appropriate location specific crops and technologies developed at the Dryland Research Station, Bankura and elsewhere in India. The strategy advocated involves (i) soil and water management, (ii) the cropping system management, (iii) fertilizer application management, and (iv) alternative land use management.

In West Bengal, about 7 per cent of the total cultivated area chronically faces different forms of land degradation and soil erosion. Soil and water conservation measures like field bunding, field levelling, bench terracing, afforestation, water harvesting structures, runoff detention tanks, sweet water ponding, etc. are being implemented by the various agencies of the State Government in areas of the State affected by soil erosion and land degradation. As a part of these conservation measures, proper utilization of scarce water and land resources in the five laterite districts of Purulia, Bankura, and Midnapore (West) and in 59 blocks of Burdwan district is being attempted through the popularization of improved dry land farming practices. This involves soil and water management, cropping system management, judicious fertilizer application and scientific land.

Minor Irrigation schemes were being implemented under different schemes like North Bengal Tarai Development Project (NBTDP), Command Area Development Programme, Swarnajayanti Gram Samridhi Yojana (SGSY), Jawahar Gram Samridhi Yojana (JGSY), Employment Assurance Scheme (EAS) and presently with NREGS.

The two-employment generation schemes, namely NREP and RLEGP were merged and a new employment generation scheme Jawahar Rojgar Yojana (JRY) was started from April 1989. JRY in the State is being implemented through the panchayats. The Centre and the State on 80:20 bases are sharing the cost of the JRY. Physical Assets like roads, tanks etc. were created under JRY. Employment Assurance Scheme (EAS) was another special employment scheme in operation in rural areas of the State. The objective of the scheme was to provide assured wage employment for 100 days during the lean agriculture season all those who are in need and seeking employment. EAS was a demand driven programme, which means that work has to be created for the rural poor seeking work. Rural infrastructure and community assets like tanks were created under EAS. The funding of the EAS has been shared by the Centre and State in the ratio of 80:20. Under the

Million Well Scheme (MWS) minor irrigation facilities were constructed. MWS later had been merged under SGSY.

It is a matter of serious concern to the problems like depletion of the ground water storage in different parts of the State, as well as the intrusion in coastal aquifers, arsenic, fluoride, iron menace in the ground water regime. As per estimate, out of 269 blocks assessed in 2008, 37 blocks fall under 'Semi-Critical' and 1 in'Critical' Blocks and the rest are in 'safe' category in West Bengal (*Economic Review, GoWB, 2009-10*). In order to regulate the utilization of ground water resources an Act titled 'West Bengal Ground Water Resource (Management, Control and Regulation) Act', 2006, was enacted and it came into force with effect from 1st August, 2006.

Attempts are also being taken to promote participatory irrigation management (PIM) of projects after completion, minor irrigation schemes being handed over to the beneficiaries committees constituted by the Panchayat Samities for operation, maintenance and management. Under Command Area Development and Water Management (CAD and WM) Programme, which is a 50:40:10 Centrally Sponsored Scheme (50% Central Share, 40% State share, 10% farmers' contribution), construction of field channels, conjunctive use of ground water and other on farm development schemes are being executed by the State Government in West Bengal.

The different programmes under which minor irrigation development in West Bengal is being currently implemented are Core Sector, Rural Infrastructure Development Fund (RIDF), Accelerated Development of Minor Irrigation (ADMI) in West Bengal, Accelerated Irrigation Benefit Programme (AIBP), Rashtriya Krishi Vikash Yojana (RKVY), Command Area Development and Water Management Programme (CADWM) etc. (*Economic Review 2009-10, GoWB*). The different programmes under which tank irrigation development may be carried out are given in appendix of the chapter.

1.4.2 District-wise Tank Irrigation Development in West Bengal

Fig. 1.3 presents percentage of tank used for irrigation purpose in West Bengal during 2000-01. As per the third minor irrigation census report out of total number of 918179 tanks in West Bengal, 161757 tanks were used for irrigation purposes in 2000-01. From Fig. 1.3, we see that higher percentage of tanks is used for irrigation purposes in the districts of Purulia, Bankura, Birbhum and 24 Parganas (South). The district-wise percentages of tanks used for irrigation purpose are

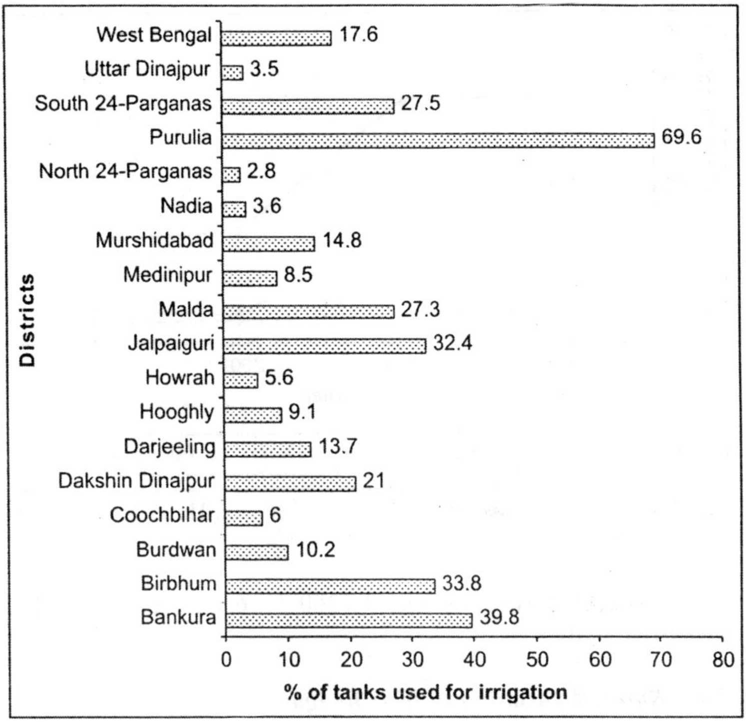

Source: GoWB, 2001, Minor Irrigation Census.

Fig. 1.3: Percentages of Tank Used for Irrigation Purpose in West Bengal

presented in the following Fig. 1.4. We see that there is variation in the percentage of tanks used for irrigation among districts.

Table 1.14 shows the district-wise area irrigated by Tanks of West Bengal during the period 2001-02 to 2005-06 (in thousand hectares). We point out in Table 1.14 that area irrigated by Tanks in Bankura, Purulia and Paschim Medinipur were 20.8, 22.4 and 26.4 thousand hectares respectively in the year 2001-02. During the above mentioned period, the tank-irrigated area decreased in Paschim Medinipur but increased in Bankura and Purulia.

The district-wise percentages of Tank irrigation over the period (1950-2006) have been given in Table 1.15. From Table, we see that the percentage of Tank irrigation falls in most of the districts including Bankura, Purulia and Medinipur districts over the years.

The graphical representation in Fig. 1.4 depicts year-wise percentages of tank irrigation of three districts in West Bengal based

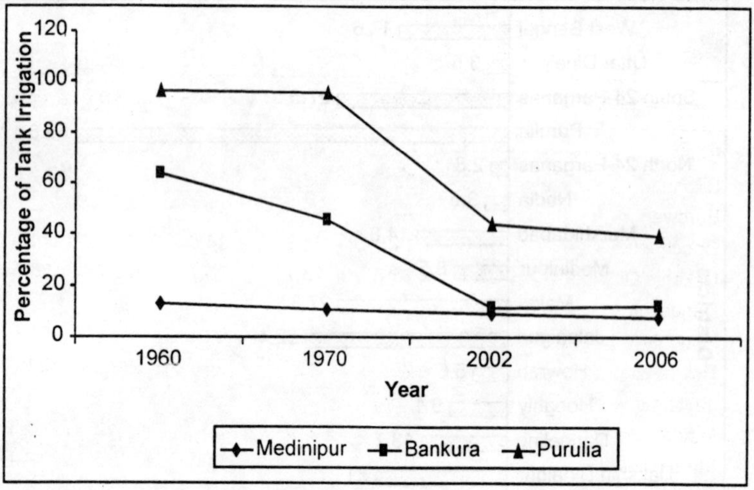

Fig. 1.4: Year-wise Percentage of Tank Irrigation in Three Districts of West Bengal

on Table 1.15 during the period 1960-2006 for the three districts under our study.

1.4.3 Rainfall Variation in West Bengal

The total water resources of a particular geographical region in a given period are the total of the quantities that enter the region: (i) From atmosphere (as precipitation), (ii) From underground (as ground water draft or discharge), and (iii) From neighbouring region (as runoff or river-flow).

West Bengal is a state of good rainfall. We have collected State-wise rainfall data for the period 1991-2006. Our calculation shows that the average rainfall of West Bengal during the period is 1612.61 mm which ranks 10th out of 36 Metrological Sub-division.

Table 1.16 shows year-wise actual annual rainfall report of different districts in West Bengal (in '000 millimetre). The average annual rainfall is higher in Medinipur district than the Purulia and Bankura districts.

District-wise co-efficeint of variation (CV) presented in Table 1.16B shows that it is high in Medinipur and Dinajpur compared to other districts. The change of second period and first period shows that it has fallen in most of the districts. Though it may not be true about intensity of rainfall.

Table 1.14: District-wise Area Irrigated by Tanks of West Bengal in the Year 2001-02 to 2006-07

('000 acres)

Districts	2001-02	2002-03	2003-04	2004-05	2005-06	2006-07
Bankura	51.4	61.5	65.9	80.0	81.8	82.7
Birbhum	100.8	100.0	62.7	62.7	62.5	62.4
Burdwan	—	—	—	—	—	—
Coochbehar	14.6	14.6	14.6	14.6	14.6	14.5
Dakshin Dinajpur	25.7	25.7	25.7	25.7	25.7	25.6
Darjeeling	.	.	.	.	.	.
Hooghly	86.9	82.3	82.5	92.6	92.6	92.5
Howrah	20.7	20.7	20.7	20.7	22.2	20.7
Jalpaiguri	5.7	5.7	5.7	5.2	5.2	5.2
Malda	3.2	3.2	3.2	3.2	3.2	3.3
Murshidabad	25.2	19.5	21.2	20.5	19.3	19.4
Nadia	—	—	—	—	—	—
North 24-Parganas	41.2	41.2	42.0	37.3	37.3	37.3
Paschim Medinipur	65.2	54.6	61.0	61.0	61.0	62.5
Purba Medinipur	98.8	65.0	65.0	65.0	65.0	64.8
Purulia	55.3	56.3	66.9	68.7	69.9	71.3
South 24-Parganas	84.7	88.2	98.8	85.7	30.4	37.0
Uttar Dinajpur	36.6	36.6	22.0	14.8	14.8	13.0
West Bengal	716.0	675.1	657.9	657.7	605.5	612.2

Source: Government of West Bengal, District Statistical Handbook, Various Issues.

The changing pattern of rainfall over the period 1980-2008 is shown in Fig. 1.5.

1.5 Literature Survey

There is vast literature about irrigation including tank irrigation development on various aspects. Some of them have been mentioned below (Table 1.17). We have classified the literature into some broad categories:

1.5.1 Importance of Tank Irrigation

Several authors have emphasized the importance of tank irrigation. We mention a few of them below.

According to Rao and Chandrakant (1984) the impact of tanks

Table 1.15: District-wise Percentage of Tank Irrigation over the Years (1950-2006)

Districts	*1950*	*1960*	*1970*	*2002*	*2006*
Purulia	—	96.78	95.61	43.7	39.2
Bankura	76.55	63.76	45.57	11.4	12.1
Birbhum	51.63	23.31	18.66	16.5	8.7
Burdwan	24.77	13.26	10.54	—	—
Medinipur	24.77	13.26	10.54	13.7	10.4
Howrah	2.68	2.04	1.96	3.8	17.4
24-Parganas	17.25	10.84	9.23	29.5	10.5
Hooghly	20.1	10.5	8.02	11.2	11.2
Nadia	14.71	5.13	2.46	—	—
Murshidabad	58.42	25.77	23.92	4.5	3.8
Malda	56.86	54	60.25	1.1	1.0
Dinajpur	38.71	48.34	49.16	14.2	20.4
Coochbehar	—	—	0.87	7.8	5.5
Jalpaiguri	1.82	1.36	0.67	2.7	2.3
Darjeeling	—	—	—	—	—
West Bengal	32.4	28.3	24.1	13.3	11.9

Source: Government of West Bengal, District Statistical Handbook, Various Issues, Chatterjee (1995).

Table 1.16A: District-wise Average annual rainfall (mm) in West Bengal for the Two Sub-periods

District	*Whole Period (1980-2008)*	*1st Period (1980-1995)*	*2nd Period (1996-2008)*	*Change*
Burdwan	1424.6	1328.4	1543.0	214.7
Birbhum	1487.2	1409.3	1583.1	173.8
Bankura	1419.5	1443.0	1390.5	–52.5
Medinipur	2405.0	1731.1	3234.3	1503.2
Howrah	1544.3	1536.5	1553.9	17.4
Hooghly	1428.2	1461.3	1387.5	–73.8
North 24 Parganas	1613.5	1735.5	1463.4	–272.1
South 24 Parganas	1861.0	1851.0	1872.6	21.6
Nadia	1419.4	1410.7	1430.1	19.4
Murshidabad	1454.1	1449.8	1459.4	9.6
Dinajpur	2828.0	2265.6	3520.2	1254.6
Malda	1610.1	1559.0	1673.0	114.0
Jalpaiguri	3515.2	3526.3	3501.5	–24.8
Darjeeling	3101.4	3041.5	3175.1	133.6
Coochbehar	3066.9	3169.0	2941.3	–227.7
Purulia	1350.1	1330.5	1374.4	43.9
West Bengal	1970.5	1890.6	2069.0	178.4

Source: Own Estimation based on Government of West Bengal, Statistical Abstract, Various Issues.

Table 1.16B: District-wise Coefficient of Variation in Rainfall in West Bengal for the Two Sub-periods

Districts	*Whole Period (1980-2008)*	*1st Period (1980-1995)*	*2nd Period (1996-2008)*	*Change*
Burdwan	19.5	16.7	19.5	2.8
Birbhum	17.7	16.4	17.6	1.1
Bankura	13.0	13.8	12.0	–1.8
Medinipur	40.7	33.0	21.3	–11.6
Howrah	16.6	18.8	14.2	–4.6
Hooghly	15.9	17.6	13.3	–4.2
North 24 Parganas	15.8	15.5	8.6	–7.0
South 24 Parganas	16.9	18.3	15.8	–2.4
Nadia	16.4	20.5	10.4	–10.1
Murshidabad	17.3	19.5	14.9	–4.6
Dinajpur	37.7	34.6	27.6	–7.0
Malda	18.3	21.4	14.0	–7.4
Jalpaiguri	16.4	17.7	15.5	–2.2
Darjeeling	13.5	13.8	13.2	–0.6
Coochbehar	17.7	18.0	17.0	–1.0
Purulia	14.8	16.1	13.6	–2.5
West Bengal	11.4	12.4	8.3	–4.1

Source: Own Estimation based on Government of West Bengal, Statistical Abstract, Various Issues.

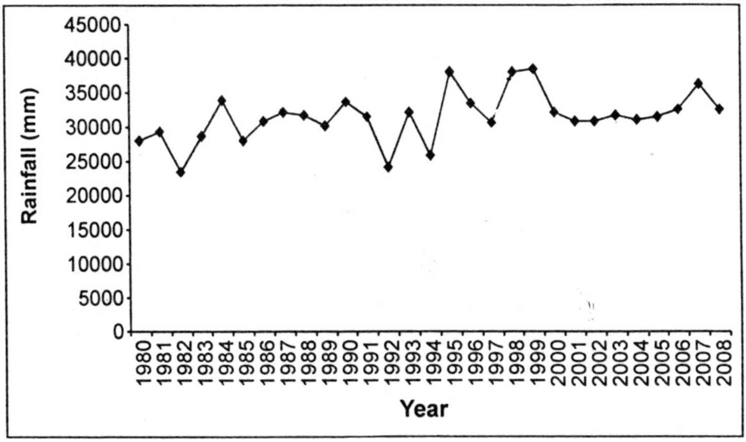

Fig. 1.5: Rainfall Pattern in West Bengal

Table 1.17: Theme-wise Literature Review

I.12.1	Importance of tank irrigation	Svasubramanyan (2006), Kannan *et al.* (1989), Rao (1978), Adams (1991), Kumar (2007), Narasaiah (2000), Mittal (2006), Banda *et al.* (2007), Rao (2005)
I.12.2	Problem areas in tank irrigation	Janakarajan (1993), Gupta (1986), Sakthivadivel (2004), Ali (1984), Moose and Reddy (2004), Reddy *et al.* (2005), Raju (2001), Joy *et al.* (2007), Goswami (2006)
I.12.3	Regional disparity in tank irrigation	Reddy (2006), Rajan (1990)
I.12.4	Innovations in tank irrigation	Pant (1998), Shah (2006), Rockstrom (2005), Easter (1977)
I.12.5	Suggestions for Improvement in Tank Irrigation	Anbumozhi *et al.* (2007), Sastry *et al.* (2003), Swaminathan (2001), Anbumozhi (2007), Sengupta (1991), Gulati *et al.* (2005), Thanh *et al.* (1990), Biswas *et al.* (1997), Sangal (1991), Kumar (2007)
I.12.6	Policy Suggestions	Vaidyanathan (2006), Sengupta (1985), Sundar (1991), Mitra (1992) Shah (2005), Kakumanu (2006), Reddy *et al.* (2006), Sreedhar (2007) Madhav Roopa (2007-08), Babu and Ramakrishna (2008), Sreedhar (2007), Palanasami and Easter (2000)
I.12.7	Efficiency	Reddy (2003), Tsur Yacov *et al.* (1997), Rydzewski (1990), Chiraneevulu (1992)
I.12.8	Institutions for Tank	Chidananda. B.L. *et al.* (1994), V.N. *et al.* (2006), Roy (2006) Maloney and Raju (1994)

would be more in the nature of stabilization rather than any substantial transformation in the economic milieu of the backward areas.

Svasubramanyan (2006) clearly points out the necessity for the revival of tanks. Wherever tank irrigation systems are available, the well water recharge is appreciably more than the independent wells' recharge through rainfall.

According to Kannan *et al.* (1989), the difference between the cost per hectare of minor and major irrigation is so large that it is doubtful whether major irrigation would turn out to be more cost-effective even after making all the allowances in favour of major irrigation.

The purpose of the note by Rao (1978) is to bring out some of the links between irrigation and development and suggest that the crucial constraints in the Indian case are likely to be institutional and socio-economic in nature.

China has been able to feed 22 per cent of the world's population with only 6.5 per cent of the world's arable land. Shi (2002) argues that Chinese ecological agriculture has significant effects on agricultural and rural development since its inception in the early 1980s.

The essay of Kumar (2007) examines various systems of water appropriation and agricultural practices in Medieval Rajasthan. The paper suggests that better management of water by peasantry along with drought resistant agricultural production has been a characteristic of pre-colonial Rajasthan.

Large-scale irrigation (LSI) has been an important feature of agricultural development investment in sub-Saharan Africa, yet LSI projects have a poor performance record. According to Adams (1991) the focus of attention has shifted to small-scale irrigation.

According to Narasaiah (2000), Irrigation has played a double role. It has not only helped raise yields through synergistic interaction with HYVs and fertilizer, but has also contributed to considerable increases in harvested area by enabling higher cropping intensity.

Mittal (2006) shows that in Shibalik region where there is no possibility of developing any other source of irrigation, rainwater harvesting has proved to be panacea. The yield of wheat through water harvesting has increased from 8 q/he to 27 q/he.

Banda *et al.* (2007) estimate the value that domestic users in the Teleport sub-basin attributable to improved quantity and reliability of water supply using contingent valuation method.

According to Sakthivadivel (2004) despite several state-sponsored tank rehabilitation programmes in many states, the number of tanks rehabilitated effectively is negligible as compared to the total number of tanks.

1.5.2 Problem Areas in Tank Irrigation

According to Janakarajan (1993), because of private control and ownership to irrigation water (viz. ground water), farmers' interest in the collective effort for maintaining traditional irrigation systems get weakened.

The Integrated Tribal Development Programme (ITDP) was formally launched in 1974 under the Fifth Five Year Plan. Dipankar Gupta (1986) discussed the incongruities in the ITDP administrative structure, the lack of control over resources, the complete absence of popular initiative, the non-involvement of popular bodies, and the complete failure of its monitoring system.

According to Ali (1984), the system of localisation followed in South India has failed in every project. Irrigation engineers generally do the design and construction of projects without involving the departments of agriculture, ground water, etc. and without taking the wishes of the farmers into consideration.

Moose and Reddy (2004) explained why did tank institutions continue to function effectively in the red soil villages while they were not effective in black soil villages. The red soil villages are dependent on water-intensive paddy crop, whereas crop diversification has taken place in black soil villages following water scarcity in earlier years.

According to Hanumantha Rao (2005), though tank irrigation based on surface water is a sustainable form of irrigation, the spurt in minor irrigation through groundwater extraction has caused a steady decline in the water table in several parts of the country.

According to Goswami (2006), the dependence on tank irrigation has declined in all the districts of West Bengal except Purulia.

It has been observed that resource poor farmers still account for major share of tank irrigation in India. Realising the importance of tank irrigation, the National Project for the repair, renovation and restoration of water bodies has been launched in March 2005 (*Ahuja 2006*). For the sustainable agricultural development in India, tank irrigation can play a very crucial role.

In recent years, the state of Andhra Pradesh in India has been a leader in economic reforms in India, particularly in the irrigation sector. The paper by Raju, K.V. (2001) compares the promise, actual practice, early results and potential implications of the reforms.

Joy *et al.* (2007) present some case studies relating to water disputes, civil disturbance caused by water shortages, misuse and mismanagement of water resources, and potential regulatory solutions to diffuse water conflicts at several levels. The case studies can be broadly categorized as: (1) Conflicts between two user groups for a common source of water, (2) Sharing of water for different purposes, (3) Inefficient water management policies, (4) Lack of monitoring, (5) Practice of conventional methods of water usage, and (6) Transgression of law.

The literature survey reveals that there is a vast literature on irrigation. However, there is very limited works on tank irrigation in West Bengal. As our study is on dry zones, we have selected three districts in West Bengal for this purpose.

Reddy *et al.* (2005) observe that devolution of powers to WUAs has not taken place as most important functions like assessment, collection of water charges, sanctioning of works etc. remain in the hands of irrigation department.

1.5.3 Regional Disparity in Tank Irrigation

Reddy (2006) focuses on irrigation development should be in terms of reducing the absolute disparities (achieving the threshold level of area under irrigation) at the district level rather than targeting the broad political regions. For, most of the public investments in the form of watershed development, tank renovation may benefit the rich if the access to groundwater is distorted.

Despite huge investments and numerous irrigation projects, Andhra Pradesh continues to face water scarcity, which results in regional disparities and political turmoil. This paper by Reddy (2006) examines the future scenario of regional distribution of irrigation in AP in the light of the past and present scenarios.

According to Sampath Rajan K. (1990), there exists considerable inequality in India in the development and distribution of irrigation in all of its aspects across rural households. Flow irrigation as a whole has higher inequality in the distribution of benefits across farm-size groups than lift irrigation as a whole in India. Of the two major contributors to the overall irrigation inequality across rural households in India, namely "between states inequality" and "within states inequality," the latter contributes more than three quarters of the total inequality. Thus, to improve equity in irrigation development and distribution more has to be done in terms of distribution across farm size groups than in terms of balanced regional irrigation development.

1.5.4 Innovations in Tank Irrigation

According to Easter (1977), the limitation of water will likely be an important restraint to adoption of new crop technologies requiring intensive use of water. Productivity increases will have to come from development and adoption of new technologies and production practices consistent with rainfed conditions.

Shah (2006) argues that Integrated Water Resources Management (IWRM) paradigm of direct demand management is not feasible in informal water economies. The rise of a class of intermediaries, i.e.

water service providers between users and natural resources of water is a precondition to meaningful demand management.

The analysis of Rockstrom (2005) shows that applying a more pragmatic development paradigm that blends top down and bottom up approaches, while integrating innovations with indigenous capacities, will generate larger opportunities for significantly improved crop yields in semi-arid and rain-fed farming system.

According to Pant (1998), the 'ahar-pyne' system in South Bihar is one such indigenous irrigation system in India, which has successfully galvanised the local farming community in large numbers against caste divisions and ecological obstacles for over 100 years and ensured equitable distribution of water among individual cultivators.

1.5.5 Suggestions for Improvement in Tank Irrigation

According to Sangal (1991), water rates of minor irrigation projects should be such that they are able to fully cover O & M charges, yearly component of loan along with its interest after accounting for subsidy for the assets created.

Sastry *et al.* (2003) opine that the productivity of dry land needs to be enhanced through watershed management approach that consisted of area planning of natural resources specially land and water. Evidence exists that supplemental irrigation increases crop production by at least 2-3 times the conventional crop production system.

Swaminathan (2001) argues dry farming areas are ideal for the cultivation of low water requiring but high value pulses and oilseeds.

Anbumozhi (2007) observes that deferred maintenance of surface irrigation infrastructure over years has led to further deterioration of its physical service.

Sengupta (1991) has emphasized the role of farmers' inherent organizing power and their imagination and local adaptability for participatory management in common property like irrigation.

Gulati *et al.* (2005) suggest that success of the participatory irrigation management depends on size of the command area, distance to market place, leadership and social capital.

Thanh *et al.* (1990) observe that as the average rate of expansion of irrigated area has been declining. Water must used efficient, i.e. in an environmentally sustainable manner. To them environmentally sound water management is an essential requirement for future development in developing countries irrigation management responsibilities should be transferred to WUAs or private/NGO contractors.

Biswas *et al.* (1997) have presented the lessons learnt from preparation and implementation of national water master plans of some countries including National Water Policy of India, 1987.

According to Anbumozhi *et al.* (2007), tank irrigation is a profitable technology in economic, environment and social terms; but under present conditions of management, it is deteriorating rapidly. In general, sustainable crop production requires better performance of these small-scale irrigation structures, i.e. tanks, which need (i) modernization of physical structures, (ii) efficient distribution of water to and in farm fields as well as among the farmers, and (iii) proper maintenance of tank system after the modernization through farmers participation.

The book by Dinesh Kumar (2007) discusses about the options that are physically and economically viable for given situations through illustrations from north Gujarat. The study argues in favour of water tariff hike by establishing that net return from farming does not suffer due to water tariff hike; instead it improves as farmers use water more efficiently or switch over to more efficient crops, provided quality irrigation has ensured.

1.5.6 Policy Suggestions

Vaidyanathan (2006) emphasizes the importance of giving far greater attention and support to improve the quality of basic information to address huge gaps in our knowledge and understanding water economy and its institutions.

According to Sengupta (1985) the knowledge of tradition is necessary to alter the distorted course of development in science and society under colonial rule.

Sundar (1991) deals with complexities of a public irrigation system, to irrigation administrators and policy planners who have to contend with a host of ideas/proposals for improving the management of public irrigation systems.

The paper by Mitra (1992) assesses comparatively the present system of operation, management and distribution of water in surface irrigation in India and the water control institutions in Japan and makes out a case for the relevance of the latter in India.

The analysis of Shah (2005) examines the economic rational for subsidies in watershed projects and highlighted the critical importance of three interrelated aspects, viz. choice of technology and treatment, link up with credit support and market and institutional mechanism for more effective allocation as well as use of subsidies.

The study by Kakumanu (2006) attempts to find a solution by employing moral suasion (creating awareness, training, educating and voluntary compliance) which can be an effective alternative approach to combat the problem of common pool resource degradation and to encourage the farmers to operate and maintain common property resources.

According to Reddy *et al.* (2006), NGO is bridging the gap in collective action in minor irrigation and improving the water productivity. By considering the modified structure of WUA, we can improve the managerial efforts for minor irrigation system.

According to Sreedhar (2007) while designing and implementing tank rehabilitation projects, it is important that these projects should made an integral part of the poverty reduction programmes of the government because the decline in collective management of tanks would adversely affect the poorer sections and he also advocates stricter rules of well irrigation should be implemented.

From case study of Madhav Roopa (2007-08), it is evident that the reforms carried out in Andhra Pradesh have dominated by a hegemonic discourse that promotes market as the solution to the problems of water resources management.

Babu and Ramakrishna (2008) have suggested measures like conservation of water, rainwater harvesting, creation of low reservoirs such as check dams and for storing the water, prevention of losses through seepage and leaks, improving usage efficiency of water, crop rotation, recycling and reuse of water.

According to Sreedhar (2007), an important challenge for tank irrigation is careful planning for crop diversification including the promotion of fodder crops, which becomes crucial for occupational diversification to promote sustainable livelihoods.

According to Palanasami and Easter (2000), government should establish demonstration tanks in different agro-climatic zones with successful experiments on tank investments as well as new management strategies.

1.5.7 Efficiency of Irrigation/Tank Irrigation

The book by Reddy (2003) gives a good insight of irrigation as a structural benefactor for analyzing productivity and efficiency.

In their article, Tsur Yacov *et al.* (1997) investigate the efficiency performance of several methods of pricing irrigation water, paying special attention to the costs associated with implementing them.

The book by Chiraneevulu (1992) discusses the problems of improvements of efficiency under conditions of tank irrigation in Andhra Pradesh and recommends for public policy for improvement of storage capacity.

According to Rydzewski (1990) small-scale irrigation, if implemented in a cost-effective manner, can show higher returns to investment than large-scale projects.

1.5.8 Institutions for Tank Irrigation

The study of Chidananda *et al.* (1994) examine the institutional factors responsible for the deterioration of tank irrigation, the community management practices for tanks and pickups and an assessment of the sustainability of the management practices towards rehabilitating the tanks for irrigation.

Using the concepts of new institutional economics, Namboodiri *et al.* (2006) have tried to analyze the performance of the local level water user associations and irrigation cooperatives in three major states, namely Andhra Pradesh, Gujarat and Maharashtra.

Maloney and Raju (1994) have discussed about different institutions of water management and have shown how farmers organize themselves to completely control and manage several types of smaller irrigation systems.

Roy (2006) has argued that watershed management programmes have frequently collapsed due to the problem of appropriate institutional structure, particularly in mobilizing stakeholders and the design of appropriate sharing mechanisms for costs and benefits. He has also cited the case of Chinese irrigation system, which is mostly based on surface water and is relatively small.

1.6 Objectives, Methodology, Study Area

1.6.1 Objectives of the Study

The main purpose of the project is to examine the prospect of tank irrigation in the dry zones of West Bengal. The following broad objectives have been taken into consideration in this study keeping in view the main purpose:

I. To find the trend of tank irrigation in West Bengal with particular reference to dry zones in West Bengal, i.e. the districts of Bankura, Paschim Medinipur and Purulia for our purpose.
II. To judge efficiency of tanks in these dry zones and to identify the factors affecting the efficiency or performance of tanks.
III. To evaluate the conditions of the tanks/bunds in the study area.
IV. To evaluate attitude study of the households about tank quality.
V. To exemplify the willingness to pay behaviour of the households for the construction/renovation of tanks.
VI. To search for the institutional pattern to involve the users of the tank.
VII. To test whether there are differences in productivity, cost and profit differences in the tank command areas.
VIII. To prescribe measures to improve the efficiency of the tanks.

1.6.2 Methodology

We have used different econometric tools and different kinds of statistical graphs, charts etc. depending upon the purpose. Some are mentioned below:

(i) We have used linear regression technique to find the determinants of productivity of tank command area.
(ii) To find the determinants of variation in proportion of non-aman (aman is the principal crop in West Bengal) cropped area to total cropped area we have used Tobit regression. It may be mentioned that proportion of non-aman total cropped area indicates the success in tank irrigation.
(iii) To test whether there is significant differences in cost per acre, production per acre and profit per acre across different households in different districts we have used ANOVA technique.
(iv) To judge the efficiency of tanks we have used data envelopment analysis (DEA) and regression techniques.
(v) For estimating total willingness to pay for renovation, we have used Contingent Valuation Method (CVM).
(vi) We have used principal component analysis for tank quality assessment by the households.

(vii) We have used cost benefit analysis technique to judge the economic viability of irrigation scheme like *happa* (small tank).

1.6.3 Study Area and Sampling Design

In order to examine the above mentioned objectives, we select the study area to suit our purpose. The precondition for dry zones in the three districts of Purulia, Bankura and Paschim Medinipur did not allow us to adopt random sampling. We used purposive sampling procedure for the selection of blocks and villages in the sample districts. The blocks have been selected with the discussion of the officials of the Departments of Irrigation, Principal Agricultural Officer and other officials in the district. The villages have been selected in consultation with GP offices. But for the selection of tanks and households in the villages we have used random sampling procedure. Table 1.18 represents the number of districts, blocks, villages and households selected in West Bengal for our survey.

Table 1.18: Number of Districts, Blocks, Villages, Tanks and Households Selected in West Bengal

Districts	*No. of blocks selected*	*No. of GPs/ Municipality selected*	*No. of villages selected*	*No. of tanks selected*	*No. of households selected*
Bankura	2	5	26	52	155
Purulia	3	8	27	60	198
Paschim Medinipur	6	14	49	69	180
Total Sample	11	27	102	181	533

Source: Based on Primary (field) Survey.

1.6.4 Design Schedule of Questionnaire

The questionnaire designed for primary survey was in two parts. One part relates to the tank level data and second part relates to the household level data. The tank schedule aims to collect information about tanks from local level representative persons. The household data was collected through personal interview method of the households. The pilot survey was conducted and the questionnaire was pretested before preparing the final questionnaire. The blocks in which the surveys had been conducted is shown in the following map.

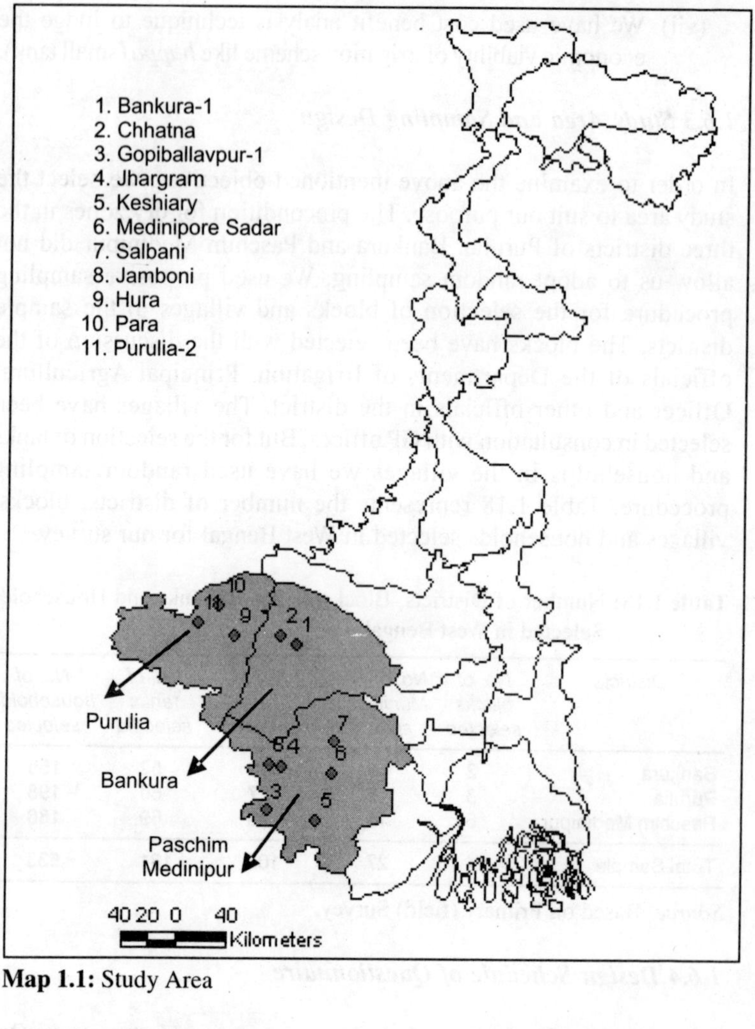

Map 1.1: Study Area

1.6.5 Data Source

The present book has used both primary and secondary data though the work is based more on primary data. The primary data used in this study have been collected using the structured questionnaire during the period of 2008-10. For secondary data we have relied on various government reports and documents like Economic Survey, Planning Commission documents, Central Water Commission

Reports, Economic Review (Government of West Bengal), Reports of Bureau of Applied Economics and Statistics—Government of West Bengal etc.

1.6.6 Schematic Approach

The entire work has been divided into six chapters to have a clear understanding of the considered objectives:

- Chapter 1 : Introduction: Irrigation Development in West Bengal with Special Reference to Tank Irrigation.
- Chapter 2 : Irrigation Development in the Districts under Study with Special Reference to Tank Irrigation.
- Chapter 3 : Tank Study based on Primary Survey.
- Chapter 4 : Household Study on Tank Irrigation based on Primary Survey.
- Chapter 5 : Some Specific Case Studies.
- Chapter 6 : Conclusion and Recommendations

Summary

In this chapter we have examined the pattern of irrigation development in West Bengal. We have seen that though in recent years the percentage share of irrigation of West Bengal in India has increased, the percentage of net irrigated area is still about 54 per cent. We have also seen the district-wise percentage of cultivable irrigated area has a wide disparity among districts. District-wise percentage share of different sources of irrigation is calculated and it shows that districts like Purulia, Bankura still depend on tank irrigation. The district-wise rainfall data have been collected for the period 1980-2008. In what follows that average rainfall has increased for the period 1996-2008 as compared to 1980-95. A literature review on tank irrigation has been presented thematically. The objectives, methodology and area of the study are also given in the chapter.

Appendix A.1.1

Water Requirement for Crops

The water requirement of different crops is presented in Table A.1.1. The water requirement for Aman is 1050 mm, for boro paddy 1500 mm etc. We see that pulses are much less water intensive compared to cereals.

Table A.1.1: Water Requirement of Different Crops (in mm)

	Crop	*Sowing season*	*Harvesting season*	*Total*
Rice	(a) Aus	May	Aug	450
	(b) Aman	July	Nov	1050
	(c) Boro	Jan	April	1500
Pulses	(a) Gram	May	March	200
	(b) Arhar	May	March	570
	(c) Mung	May	March	220
	(d) Musur	May	March	200
	(e) Khesari	May	March	220
	(f) Other pulses	May	March	220
	Miscellaneous pulses	May	March	200
Vegetables	(a) Potato	Nov	Feb to March	425
	(b) Dry Chillies	April (year round)		370
	(c) Ginger (Dry)	December (Year round)		370
Oilseeds	(a) Rapeseed and Mustard	Nov	Feb	400
	(b) Linseed	November	Feb	170
	(c) Seasame	Feb-June (1st cycle), Oct-Dec (2nd cycle)		500
	(d) Other oilseed(s)	Nov	Feb	400
Cereals	(a) Wheat	Nov	March	420
	(b) Barley	Nov	March	300
	(c) Maize	Nov	March	550
	(d) Jowar	Nov	March	400
	(e) Bajra	Nov	March	400
	(f) Ragi	Nov	March	400
	(g) Small Millets	Nov	March	400
Commercial crops	(a) Jute	May	Aug	400
	(b) Mesta	April	Aug to Oct	500
	(c) Cotton	April	Oct	1200
	(d) Sunhemp (Fibre)	April	Oct	400
	(e) Tea	Year round		200
	(f) Sugarcane	Dec	Nov	2500
	(g) Tobacco	Nov	March	600

Source: West Bengal Pollution Control Board (WBPCB), 2009.

Appendix A.1.2

There are different schemes for tank irrigation development in India. Some of them are given below:

(1) *Swarnajayanti Gram Swarozgar Yojana (SGSY)*: SGSY has been introduced from 1st April, 1999 by replacing IRDP and its programme like TRYSEM, DWCRA, SITRA, GKY, MWS with source of fund from Central and State in 75:25 ratio.

(2) *Rashtriya Sama Vikas Yojana (RSVY)*: RSVY, a centrally sponsored programme for the opportunities available for development in the district and to fill up critical gaps for enhancing the productivity and employment.

(3) *Modified Accelerated Irrigation Benefits Programme (AIBP) 2006*: Under this Scheme for non-special category states, only those minor irrigation schemes with potential more than 50 hectare, which serve tribal areas and drought prone areas, could be included under AIBP.

(4) *National Rural Employment Guarantee Act (NREGA)*: Providing not less than one hundred days of guaranteed employment, the programme includes water conservation and water harvesting in the form of excavation and construction of tanks, check dams, percolation tanks, underground dykes, ponds, rain water-harvesting structures and drought proofing including all the components of watershed development.

(5) *Rashtriya Krishi Vikas Yojana (RKVY), 2007*: This scheme is being implemented under Department of Agriculture and Co-operation. The RKVY is a State Plan Scheme and seeks to encourage convergence with schemes like NREGS, SGSY and BRGF, Development of rainfed farming systems in and outside watershed areas, as also integrated development of watershed areas, wastelands, river valleys.

(6) *Common Watershed Development Projects (CWDP), 2008*: In co-ordination witn the Planning Commission, an initiative has been taken to formulate "Common Guidelines for Watershed Development Projects" in order to have a unified perspective by all ministries. States will now be empowered to sanction and oversee the implementation of watershed

projects within their areas of jurisdiction and within the parameters set out in these guidelines.

(7) *Joint Convergence Guideline*: National Rural Employment Guarantee Act (NREGA) (Ministry of Rural Development) and Programmes of Water Resources (Ministry of Water Resources): The Ministry of Water Resources, which has represented on the Task Force, has identified as an important partner to NREGA, as more than 50 per cent of NREGA works relate to water conservation/management.

(8) *Accelerated Irrigation Benefits Programme (AIBP)*: This programme is for Major, Medium and Minor Irrigation Projects and Extension, Renovation and Modernization of Irrigation Projects which having investment clearance of Planning Commission can be completed in the next four financial years.

(9) *Farmer's Participatory Action Research Programme (FPARP)*: Under this programme new irrigation technologies developed by various agricultural universities, research institutes, etc. are being demonstrated to farmers as a measure of creating awareness amongst farmers to save water.

(10) *Command Area Development and Water Management Programme (CADWM)*: Command area development and water management programme aims at efficient utilisation of irrigation water for optimizing agricultural productivity and production from irrigated lands in a co-ordinated manner on a sustainable basis and increasing the pace of tackling the problem of water-logging by bringing in newer technologies like sub-surface drainage and bio-drainage under the realm of this programme.

(11) *Repair, Renovation and Restoration (RRR) of Water Bodies, 2005*: The Scheme of RRR of Water Bodies includes the comprehensive improvement of selected tank systems, catchments area treatment, command area development and capacity building of stakeholders.

(12) *Hariyali (2003)*: This involves village communities in the implementation of watershed projects under all the area development programmes, namely Integrated Wastelands Development Programme (IWDP), Drought Prone Areas Programme (DPAP) and Desert Development Programme (DDP).

(13) *National Bank for Agriculture and Rural Development (NABARD)—Watershed Development Fund, 2006*: The Union Finance Minister, in his Budget Speech for 1999-2000 had announced the creation of a Watershed Development Fund (WDF) with the National Bank for Agriculture and Rural Development (NABARD) with broad objectives of unification of multiplicity of watershed development programmes into a single national integrated programme.

(14) *National Watershed Development Project for Rainfed Areas (NWDPRA)*: It is a centrally sponsored scheme. The twin objectives of the schemes are sustainable production of biomass and restoration of ecological balance in the watershed areas.

(15) *Drought-Prone Area Programme (DPAP)*: The programme is being implemented with the prime objective of promoting economic development of the watershed community by putting the natural resources like land water etc. to the optimum use.

Appendix A.1.3

The Bengal Tanks Improvement Act, 1939. (As modified up to the 1st April, 1967)

This Act had been enacted to provide for the improvement of tanks in Bengal for purposes of irrigation. According to this Act if the Collector is of opinion that any tank has fallen into disrepair or disuse, he may serve a notice in the prescribed form and manner on the person having control over the tank to carry out within a period specified in the notices such improvements of the tank as the Collector considers necessary for the proper utilization of the tanks for the purposes of irrigation.

If the person having control over the tank does not send any intimation or send intimation to the Collector within the period fixed under norms that means he is not willing to carry the improvements, then the Collector may declare the tank to be a derelict irrigation work.

Section 5 of the Tank Improvement Act states the following:

After the notice, declaring a tank to be a derelict irrigation work has been confirmed under Section 4 the Collector, if he thinks fit, may at any time—

(a) take possession of the tank and carry out the improvements specified in the notice under Section 3, or
(b) authorize under Section 6 a local authority, co-operative society, or any other person interested to take such action.

Chapter 2

Irrigation Development in the Districts under Study with Special Reference to Tank Irrigation

In the previous chapter, we have discussed the irrigation development in West Bengal with particular reference to tank irrigation. Now, in Chapter 2, we will discuss the irrigation development of three districts under our study in West Bengal, *viz*. Bankura, Purulia and Paschim Medinipur with special emphasis on tank irrigation. As our study area falls in these three districts, understanding the nature and pattern of irrigation particularly tank irrigation development is very important. This chapter includes administrative profiles, demographic variation, physical characteristics, irrigation with particular emphasis on tank irrigation and cropping pattern in the districts. Before going into the reports of Primary Survey we shall highlight these in this chapter.

2.1 Administrative Profiles of the Districts

In this section we will give brief administrative profiles of the districts.

Bankura: Bankura, the fourth largest district of West Bengal is located in the western part of the state, which is popularly known as 'Rarh' from time immemorial. It has an area of 6,882 sq. km. and according to the Census of 2001, contains a population of 31.92 lakh. As Census 2001 data reveal, it ranks 12th in the state as far as total population is concerned. Damodar, Darkeswar and Kangsbati drain Bankura along with their tributaries of which Gandeswari, Silai and Kumari deserve separate mention. This district has a moderate deposit of Coal and a number of good deposits of China Clay.

Paschim Medinipur: Paschim Medinipur district has an area of 9,786 sq. km. and according to the Census of 2001, it has a population of 51.93 lakh. Net cropped area of the district is about 5.8 lakh he. It is bounded by Purba Medinipur district in the east and south and Bankura district in the west. Subarnarekha and Kasai River drain Paschim Medinipur district. More than 80 per cent of total population in the district is residing in the villages, and rural economy mainly depends on agriculture and agro-based nature of small industries. The District has variegated soil type, climate, eco-morphology accompanied by varieties of population, caste, creed etc. The climate is tropical and land surface is characteristic by hard rock upland, lateritic covered, flat alluvial and deltaic plans. River Kangasabati, Silaboti, Dulong, Subarnarekha etc. are flowing through this District and enriched cropping system.

Purulia: Purulia, the westernmost district of West Bengal, is well known as a drought prone district and falls within the semi-arid region of the state. Cultivation of this district is predominantly mono-cropped. Out of total geographical land, 52.47 per cent are used for agriculture. 29.69 per cent are under forest coverage (including social forestry) and 10.15 per cent are identified as wasteland. Purulia district is part of the Chhotanagpur Plateau in India. The total geographical area of the district is 6259 sq. km (*Census* 2001). Purulia is drained by Kasai, Kumari etc. with their few tributaries. Tribes of Purulia are mostly a rural phenomenon, almost 95 per cent of tribal communities habited in villages. At present, the district comprises 20 CD Blocks, 170 Gram Panchayats, 2683 Mouzas and 2468 inhabited villages. Administrative profiles of the three districts are given in the following Table 2.1.

Table 2.1: Administrative Profiles of Bankura, Paschim Medinipore and Purulia District

	Bankura	Paschim Medinipur	Purulia
Sub-division	3	4	3
Block	22	29	20
Municipality	3	8	3
Gram-Panchayat	190	290	170
Village	5187	7581 (Inhabited)	2468
Police Station	22	27	

Source: District Statistical Hand Book, Government of West Bengal.

2.2 Demographic Features of the Districts

Bankura: The percentage of Scheduled Caste population in the district is 31 per cent and the percentage of ST population in the district is 14 per cent. There is huge discrepancy between the male and female literacy rate. The percentage of non-workers is as high as 59 per cent. The incidence of poverty in Bankura is 41.5 per cent, which is much higher than the national average of 26.1 per cent in 1999-2000. The Western and the Southern parts of the district are backward compared to other two parts. Continuous degradation of natural resources, severe soil erosion, depletion of ground-water reserves, low agricultural productivity, low wage rate etc. are some of the endemic problems in these areas. These are the most important challenges in front of the district.

Paschim Medinipur: As per Census 2001, the total population in Purulia was 25 lakh. Incidence of poverty in Paschim Medinipur at 33.9 per cent is still much above the national (26.10% in 1999-2000) and the state level (27.02% in 1999-2000). The poverty in Paschim Medinipur district is more concentrated in the western parts of the district.

Purulia: The landscape of Purulia is quite unique having distinctive soil, terrain, drainage, forest and whole gamut of physical environmental set-up which is within the ambit of Rarh Tract of West Bengal. According to the 2001 Census the district contains population of 25.36 lakh. Demographic features of the Purulia district are summarized in Table 3.41. Purulia district is having second highest percentage of tribal population (18.3%) after Jalpaiguri district (18.9%). The percentage of BPL families in Purulia is 32.9 per cent and block-wise figures reveal that it is high in the blocks like Bandowan, Manbazar.

Some comparative features of the three districts are presented in Table 2.2.

2.3 Land Types in the Districts

Bankura: This district forms an intermediate tract lying between the rice producing alluvial plains of West Bengal in the east. Along the middle of the district, the ground-surface rises gradually in undulating plains. Moreover, as the land is undulated, laterite and porous sub-soil usually contains very poor moisture, erratic nature of rainfall

Table 2.2: Demographic Features of Bankura, Paschim Medinipore, Purulia (As per Census 2001)

	Bankura	Paschim Medinipur	Purulia
Male (%)	51.2	50.99	51.18
Female (%)	48.8	49.01	48.82
Scheduled Caste (%)	31.2	18.8	19.35
Scheduled Tribe (%)	10.4	16.4	19.22
Population between 0-6 age group (%)	14.1	41.04	–
Main Workers (%)	29.6	–	25.43
Marginal workers (%)	15.1	–	19.03
Decennial growth rate 1991-2001 (%)	13.8	–	14.02
Male Literacy Rate (%)	77	81.30	73.70
Female Literacy Rate (%)	49	59.10	36.50
Rural Population (%)	92.6	88.10	89.93
Urban Population (%)	7.4	11.90	10.07
Population Density (per square km)	464	531	405
Total Population (Numbers)	31,92,695	51,93,411	25,36,516
% of BPL	41.5	33.9	32.9

Source: Own Estimation based on District Statistical Hand Book. Government of West Bengal

becomes a potential threat to the crops. Fluctuating rainfall with intermittent dry spell between two successive rainfalls makes the crop, generally the Kharif crop, very vulnerable and affects the yield seriously. In case the drought spell lengthens, it takes a heavy toll on the crop resulting in low output of grain and low income. Mainly three types of lands are seen in the district:

(i) *Patit lands*: These lands with rock outcrops and *cankars* are so difficult for agriculture that they are either left fallow or are heavily grazed.

(ii) *Baid lands*: These lands occur at a level lower than that of the *Patit lands*. It is covered with sandy loam and that is given to paddy cultivation but the yield is very low.

(iii) *Kanali lands*: These lands are found at a still lower level and, with an assured moisture supply. The average yield is better than that of *Baid land*.

(iv) *Sol or Bahal lands*: In between *Kanali* and marshy area, where the water table is higher and the surface moisture ample, occurs this type of land with loamy moist or clayey soils suitable for cultivation of paddy or summer vegetables.

Purulia: Types of lands in Purulia are also classified in four categories as in Bankura namely—Patit, Baid, Kanali and Bahal lands.

Paschim Medinipur: More than 3 lakh hectares (nearly 36% of the total geographical area of District) of the district is identified as disadvantageous area and this requires soil conservation measures. A good amount of land area (nearly 48 thousand he.) is not under use now. The land type in the district is as follows: 40 per cent of the area in Paschim Medinipur is high land.

2.4 Farmers' Land Holding and Paddy Productivity in the Districts

In Bankura 65 per cent of total number of farmers are marginal with average land holding size of 0.52 he and their gross holding area is about 32 per cent. In Paschim Medinipur, 84.9 per cent of total number of farmers are marginal with average land holding size of 0.52 he and their gross holding area is about 61 per cent. In Purulia, 70.7 per cent of total numbers of farmers is marginal with percentage of land area 38.5 per cent of total land. 21.1 per cent of total number of farmers is small with percentage of land holding of 34.4 per cent. So, more than 90 per cent farmers belong to the category of marginal and small farmers with land holding of about 72 per cent.

Table 2.3: Land Class-wise Percentage of Different Land Holdings and Distribution of Area in Bankura, Paschim Medinipore, Purulia District

Type of Farmers	*Bankura*		*Paschim Medinipur*		*Purulia*	
	No. of holding	*Area*	*No. of holding*	*Area*	*No. of holding*	*Area*
	in %	*in %*	*in %*	*in %*	*in %*	*in %*
Marginal	65	32	84.9	61.0	70.7	38.5
Small	23	34	12.1	27.2	21.1	34.4
Semi-medium	10	26	2.8	10.3	7.4	22.5
Medium	2	8	0.2	1.4	0.7	4.3
Large			0.003	0.1	0.009	0.3

Source: Own Estimation based on *District Statistical Hand Book*, Government of West Bengal.

2.5 Soil Types in the Districts

Bankura: The district of Bankura is divided into three distinct topographical regions.

(i) *Western part*: This part is hilly terrain, which is characterized by large granite rocks. Its undulating terrain is susceptible to soil erosion and the hard granite base leaves lesser scope for economic production.

(ii) *Central part*: A major part of the land in the central region is lateritic in nature and the soil is devoid of many basic nutrients essential for normal crop.

(iii) *Eastern part*: It is characterized alluvial and loamy soil. The acidic nature of the soil reduces the efficacy of fertilizers. Moreover, the land being undulating, lateritic and porous, results in poor subsoil moisture which becomes a potent threat to the crop. Fluctuating rainfall, with intermittent drought spell between two successive rainfalls, makes the crop, generally the Kharif crop, very vulnerable and seriously affects the yield.

Purulia: Due to several decades of non-sustainable land use practices and highly erosive monsoon rains, deforestation and accelerated soil erosion are proceeding rapidly in many parts of the plateau. In between *Kanali* and marshy area, where the water table is higher and the surface moisture ample, occurs *Sol or Bahal* type of land with loamy moist or clayey soils suitable for cultivation of paddy or summer vegetables.

Paschim Medinipur: It shows that the sandy loam is looking more than other soil types in Paschim Medinipur district. In Paschim Medinipur district soil type is mainly lateritic and alluvium.

2.6 Ground Water Conditions in the Districts

Bankura: The water table in Bankura generally declines with varying gradients from west, northwest to east and southeast direction and broadly conforms to the topographical slope. The western sector, comprising mainly crystalline rock ground water occurs in the weathered mantle of varying thickness from 6 mts. to 15 mts. In many parts of the western area lateritic gravel lies on the weathered basement, rock that attributes to the favourable condition for percolation of rainwater. In the middle sector covered by laterite and older alluvium, ground water occurs in the moderately thick to thin aquifer where the ground water is mainly extracted through open dug well with limited number of low-duty tube wells. In the eastern alluvial area as in Indus, Kotalpur and Joypur Blocks ground water occurs under confined condition below a blanket of clay whose thickness varies around 10 mts.

Paschim Medinipur: Both ground water and surface water are being harnessed for implementation of minor irrigation schemes in various parts of Paschim Medinipur district. The district has widely varied geo-morphological agro-climatic conditions. The eastern and south-eastern parts of the district are underlain by recent alluvium except Garhbeta, Salbani, and parts of Chandrakona and Keshpur blocks.

A comparative study in the blocks is done by SWIDD shows ground water variation. It appears that 10 years average ground water varies from 8.91 mbgl. to 3.76 mbgl. during pre-monsoon [April – May] and post-monsoon [November – December] and in 2004 average ground water varies from 9.76 mbgl. to 3.49 mbgl. From comparative study it reveals that in pre-monsoon groundwater is decreasing by (-) 0.85 mbgl. and in post-monsoon it is increasing by 0.27 mbgl. As a whole, the ground water potential is decreasing in the district and in different blocks under Midnapore, Kharagpur and Jhurgram Sub-division it is alarming condition.

Table 2.4: Ground Water during Pre-monsoon, Post-monsoon and Deviation

Season	*10 year average [mbgl.]*	*Average in 2004 [mbgl.]*	*Deviation from average [mbgl.]*
Pre-monsoon	8.91	9.76	–0.85
Post-monsoon	3.76	3.49	0.27

Mbgl. — metre below ground level

Source: Government of West Bengal, District Planning Office, Paschim Medinipur.

2.7 Rainfall and Temperature in the Districts

Rainfall is a very important factor in tank irrigation. Rainfall pattern in the districts is given in Table 2.5. The average annual rainfall in

Paschim Medinipur is higher than the districts of Bankura and Purulia. As the rivers in Bankura and Purulia are mostly seasonal and remain practically dry over the major part of a year, there is no assured supply of water from this source for cultivation. The districts receive an annual rainfall ranging between 1200-1500 mm. Though this quantum of rainfall is generally sufficient, crops are liable to fail if the distribution of rain is not uniform. The greatest damage is caused if rainfall is insufficient in the months of September and October, when a good supply of water is needed for grain filling. As there is high rainfall in the monsoon months, there is scope of conservation of water for the dry months. Practicality, the districts suffers from the vagaries of rainfall and in most of the years; there has been distress in some blocks.

Table 2.5: Month and District-wise Rainfall (1980-95, 1996-2005)

(Millimetre)

Month	*Bankura*		*Medinipur*		*Purulia*	
	1980-1995	*1996-2005*	*1980-1995*	*1996-2005*	*1980-1995*	*1996-2005*
January	9.2	17.2	13.2	26.7	12.3	15.5
February	24.6	22.1	23.2	31.6	35.0	21.6
March	33.2	30.9	22.8	88.7	45.4	26.2
April	40.9	50.1	31.8	91.9	56.7	23
May	104.3	91.5	66.7	192.3	156.3	81.9
June	251.4	244.8	240.0	430.4	312.7	262.7
July	316.9	281.2	312.7	516	341.6	282.9
August	302.6	273.8	273.3	554.6	338.9	277.2
September	229.2	225.7	239.1	469	277.7	205.9
October	75.9	129.5	80.2	381.9	108.3	115.2
November	24.6	10.8	22.1	35.1	44.3	8.1
December	14.3	6.4	15.3	10.8	8.7	6.6
Annual Total	1443.0	1385	1330.5	2829	1737.8	1332.2

Source: Own Estimation based on District Statistical Hand Book, Government of West Bengal.

Annual rainfall in three districts is shown in Fig. 2.1. As Table 2.5 shows annual rainfall pattern in the three districts is more or less same except in later years. Paschim Medinipur district has experienced higher fluctuation.

According to IMD, if the annual or seasonal CV (co-efficient of variation) is 30 per cent or more, the rainfall is said to be of erratic nature and the region is drought prone (*Chatterjee 1995*). High CV

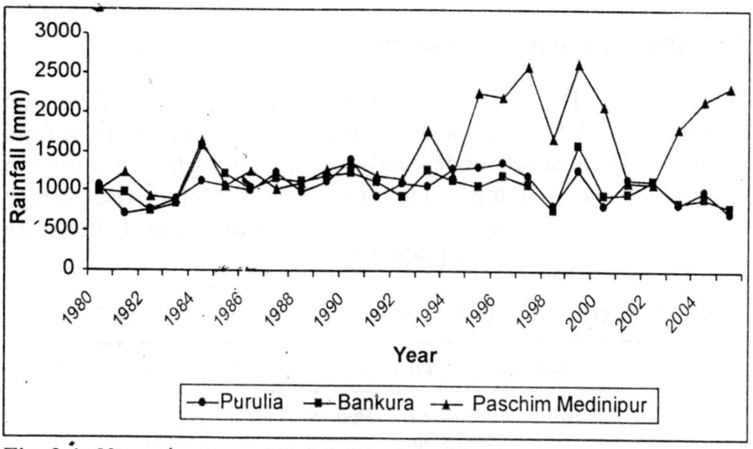

Fig. 2.1: Year-wise Annual Rainfall in three Districts

value implies poor reliability on rainfall for agriculture. The coefficient of variation for the districts during the period of 1980-2005 is given in the following Table 2.6. It should be noted that co-efficient of variation in the winter season is high in all the districts. It is also to be noted that coefficient of variation of Paschim Medinipur district is higher compared to other districts.

Table 2.6: Month and District-wise Rainfall 1980-2005: Co-efficient of Variation (CV)

Month	Bankura	Paschim Medinipur	Purulia
January	140.7	101.6	140.4
February	98.2	79.0	109.3
March	88.6	104.9	111.3
April	57.0	74.0	65.2
May	49.0	73.6	55.0
June	37.6	38.2	45.3
July	27.0	37.6	28.0
August	35.0	50.1	25.9
September	43.5	59.0	48.5
October	67.8	120.0	71.7
November	215.7	142.9	180.0
December	202.6	107.8	186.5
Annual Total	14.5	39.3	14.9

Source: Own Estimation based on District Statistical Hand Book. Government of West Bengal.

2.8 Temperature in the Districts

Temperature also plays a very important role in tank irrigation development. A high temperature leads to higher evaporation from the tanks. The districts under study face a very high temperature in the summer season compared to other districts. Month-wise average maximum and average minimum temperature in the districts are presented in the following Table 2.7.

Table 2.7: Average Temperature of Three Districts in Period 1996-2007

Month	*Maximum Temperature*			*Minimum Temperature*		
	Bankura	*Paschim Medinipur*	*Purulia*	*Bankura*	*Paschim Medinipur*	*Purulia*
January	24.5	24.5	22.8	11.8	13.6	11.5
February	28.8	27.8	23.8	15.8	17.4	15.5
March	33.2	32.5	29.0	20.5	21.5	20.0
April	36.1	34.6	32.3	24.6	24.5	24.0
May	36.8	35.9	33.8	26.8	26.3	26.3
June	34.3	33.5	32.3	26.8	26.3	26.3
July	32.2	31.7	29.8	26.3	26.3	25.8
August	31.8	30.7	29.3	26.4	26.1	25.8
September	32.3	31.5	31.0	25.9	25.7	25.0
October	31.4	31.0	30.0	23.1	23.5	22.8
November	29.5	28.8	27.8	17.8	18.5	18.5
December	25.8	25.4	23.8	12.9	14.5	13.8

Source: Own Estimation based on District Statistical Hand Book. Government of West Bengal.

2.9 Irrigation Development in the Districts

In this sub-section we present source-wise irrigation development in the districts under study, i.e. Bankura, Paschim Medinipur and Purulia.

Bankura

Source-wise irrigation development in the district of Bankura for the period of 1996-97 to 2005-06 is presented in Table 2.8. As seen in the Table, the tank irrigated area has decreased from 50.4 thousand hectares to 33.1 thousand hectares during the above mentioned period.

Table 2.8: The source-wise Irrigated Area (he) in Bankura during 1996-97 to 2005-06

Year	*Canal*	*Tank*	*DTW*	*STW*	*RLI*	*ODW*	*Others*	*Total*
1996-97	129896	50401	9510	93156	8370	0	20478	3,11,811
1997-98	178482	52968	10005	56581	8675	0	20275	3,26,986
1998-99	160819	44741	16342	97265	7892	0	16397	3,43,456
1999-00	178097	44600	17033	99429	8339	0	17467	3,64,965
2000-01	99392	39478	19293	102286	7803	0	12907	2,81,159
2001-02	142943	35931	23466	100196	5946	0	10871	3,19,353
2002-03	152732	40464	22332	98105	3814	5864	8835	3,32,146
2003-04	149407	42206	2653	106412	5455	5637	11047	3,22,817
2004-05	109890	32350	2628	47985	7252	2167	6316	2,08,588
2005-06	109624	33105	2628	47433	5375	2520	6196	2,06,881

Source: Own Estimation based on District Statistical Hand Book, Government of West Bengal.

Source-wise percentage of irrigated area of Bankura for the period of 1996-97 to 2005-06 is presented in the following Table 2.8.1. As seen in the Table, the percentage of tank irrigated area has a decreasing tendency during the above mentioned period.

Table 2.8.1: Source-wise Percentage Irrigated Area in Bankura during 1996-97 – 2005-06

Year	*Canal*	*Tank*	*DTW*	*STW*	*RLI*	*ODW*	*Others*	*Total*
1996-97	41.7	16.2	3.0	29.9	2.7	0.0	6.6	100.0
1997-98	54.6	16.2	3.1	17.3	2.7	0.0	6.2	100.0
1998-99	46.8	13.0	4.8	28.3	2.3	0.0	4.8	100.0
1999-00	48.8	12.2	4.7	27.2	2.3	0.0	4.8	100.0
2000-01	35.4	14.0	6.9	36.4	2.8	0.0	4.6	100.0
2001-02	44.8	11.3	7.3	31.4	1.9	0.0	3.4	100.0
2002-03	46.0	12.2	6.7	29.5	1.1	1.8	2.7	100.0
2003-04	46.3	13.1	0.8	33.0	1.7	1.7	3.4	100.0
2004-05	52.7	15.5	1.3	23.0	3.5	1.0	3.0	100.0
2005-06	53.0	16.0	1.3	22.9	2.6	1.2	3.0	100.0

Source: Own Estimation based on District Statistical Hand Book, Government of West Bengal.

Source-wise percentages for different sources of irrigation for the district of Bankura are shown in the following figure showing the importance of tank irrigation.

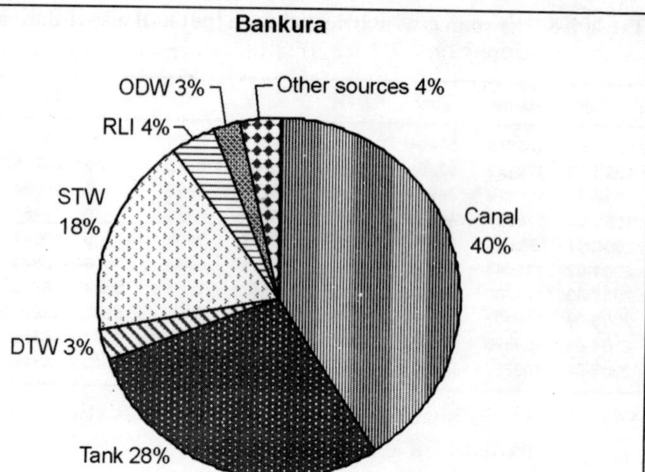

Fig. 2.2: Irrigation Source-wise Percentage of Irrigated Area in Bankura for 2001-2005

Paschim Medinipur

With all the available irrigation sources like DTW, STW, RLI, Ponds and other sources and irrigation system like Kangsabati project, approximately 63 per cent of the net-cropped area is irrigated in the district of Paschim Medinipur. The experience of drought and flood area is common in a district, which occurs off and on, and another floods sometime one. As Table 2.9 shows tank irrigated area has decreased from 3.09 thousand hectares to 2.46 thousand hectares.

Table 2.9: Source-wise Irrigation (ha) Development in Paschim Medinipur

Year	*Canal*	*Tank*	*DTW*	*STW*	*RLI*	*ODW*	*Others*	*Total*
1996-97	23115	30959	8923	93926	11945	0	89143	258011
1997-98	27302	30155	8426	84493	14558	0	49390	214324
1999-00	67274	33461	9793	123542	14139	0	49551	297760
2000-01	67274	33461	9813	116009	14129	0	50151	290837
2001-02	55296	26376	28970	109604	11898	8350	44020	284514
2002-03	62125	22089	27144	132388	16515	9636	29275	299172
2003-04	65305	24698	27144	112926	15690	9912	37699	293374
2004-05	66105	24698	32144	112926	15690	9912	37699	299174
2005-06	66105	24698	32144	110573	15690	9852	42549	301611

Source: Own Estimation based on District Statistical Hand Book. Government of West Bengal.

Source-wise percentage of irrigated area of Paschim Medinipur for the period of 1996-97 to 2005-06 is presented in the following table. As seen in the Table, the percentage of tank irrigated area has decreased during the above mentioned period.

Table 2.9.1: Source-wise Percentage of Irrigated Area during 1996-97 – 2005-06 in Paschim Medinipur District

Year	*Canal*	*Tank*	*DTW*	*STW*	*RLI*	*ODW*	*Others*	*Total*
1996-97	9.0	12.0	3.5	36.4	4.6	0.0	34.6	100.0
1997-98	12.7	14.1	3.9	39.4	6.8	0.0	23.0	100.0
1999-00	22.6	11.2	3.3	41.5	4.7	0.0	16.6	100.0
2000-01	23.1	11.5	3.4	39.9	4.9	0.0	17.2	100.0
2001-02	19.4	9.3	10.2	38.5	4.2	2.9	15.5	100.0
2002-03	20.8	7.4	9.1	44.3	5.5	3.2	9.8	100.0
2003-04	22.3	8.4	9.3	38.5	5.3	3.4	12.9	100.0
2004-05	22.1	8.3	10.7	37.7	5.2	3.3	12.6	100.0
2005-06	21.9	8.2	10.7	36.7	5.2	3.3	14.1	100.0

Source: Own Estimation based on District Statistical Hand Book, Government of West Bengal.

Source-wise percentages for different sources of irrigation for the district of Paschim Medinipur for the sub-period 2001-05 are shown in the following figure showing the importance of tube well irrigation in the district.

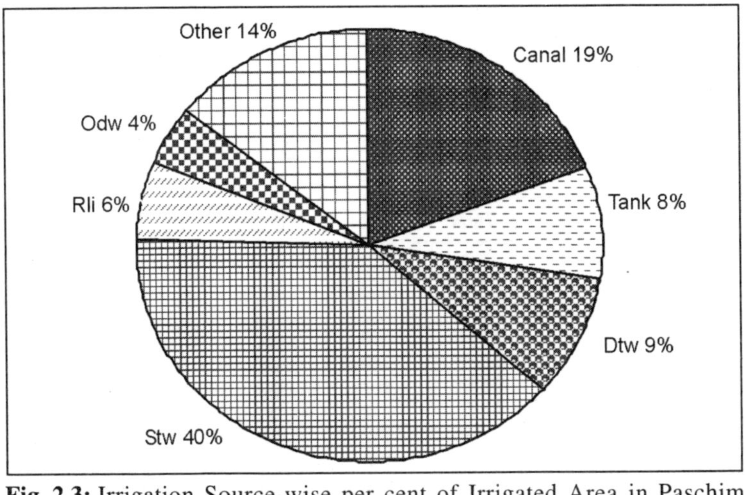

Fig. 2.3: Irrigation Source-wise per cent of Irrigated Area in Paschim Medinipur for 2001-2005

Purulia

Source-wise irrigation development in the district of Bankura for the period of 1996-97 to 2005-06 is presented in the following table. As seen in the Table, the tank irrigated area has remained stable during the above mentioned period. It was 84.66 thousand hectares in 1996-97 and 84.946 thousand hectares in 2005-06.

Table 2.10: Source-wise Irrigation (ha) Development in Purulia

Year	Canal	Tank	RLI	ODW	Others	Total
1996-97	44865	84660	3989	0	43332	176846
1997-98	46768	87543	3193	0	6557	143488
1998-99	47545	90405	5258	0	19109	162317
1999-00	55196	105662	5086	0	18548	184492
2000-01	50423	98915	1392	0	13729	164459
2001-02	52863	98435	1041	1090	2928	156357
2002-03	38518	76474	1432	1020	2963	120407
2003-04	49627	98434	1115	3994	4594	157764
2004-05	49871	98892	935	4045	3186	156929
2005-06	44844	84946	1323	3418	8823	143354

Source: Own Estimation based on District Statistical Hand Book, Government of West Bengal.

Source-wise percentage of irrigation area of Bankura for the period of 1996-97 to 2005-06 is presented in the following Table. As seen in the Table, the percentage of tank irrigated area is more or less stable.

Table 2.10.1: Source-wise Percentage of Irrigated Area during 1996-97 – 2005-06 in Purulia District

Year	Canal	Tank	RLI	ODW	Others	Total
1996-97	25.4	47.9	2.3	0.0	24.5	100.0
1997-98	32.2	61.0	2.2	0.0	4.6	100.0
1998-99	29.3	55.7	3.2	0.0	11.8	100.0
1999-00	29.9	57.3	2.8	0.0	10.1	100.0
2000-01	30.7	60.1	0.8	0.0	8.3	100.0
2001-02	33.8	63.0	0.7	0.7	1.9	100.0
2002-03	32.0	63.5	1.2	0.8	2.5	100.0
2003-04	31.5	62.4	0.7	2.5	2.9	100.0
2004-05	31.8	63.0	0.6	2.6	2.0	100.0
2005-06	31.3	59.3	0.9	2.4	6.2	100.0

Source: Own Estimation based on District Statistical Hand Book, Government of West Bengal.

Source-wise percentages for different sources of irrigation for the district of Purulia for the sub-period 2001-2005 are shown in the following figure showing the importance of tube well irrigation in the district.

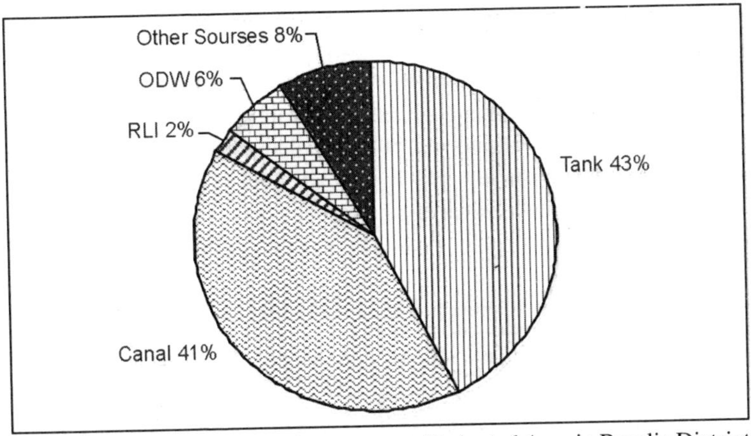

Fig. 2.4: Irrigation Source-wise per cent of Irrigated Area in Purulia District for 2001-2005

The percentage of tank irrigated area for the three districts over the period 1996-97 to 2005-06 is shown in the following figure.

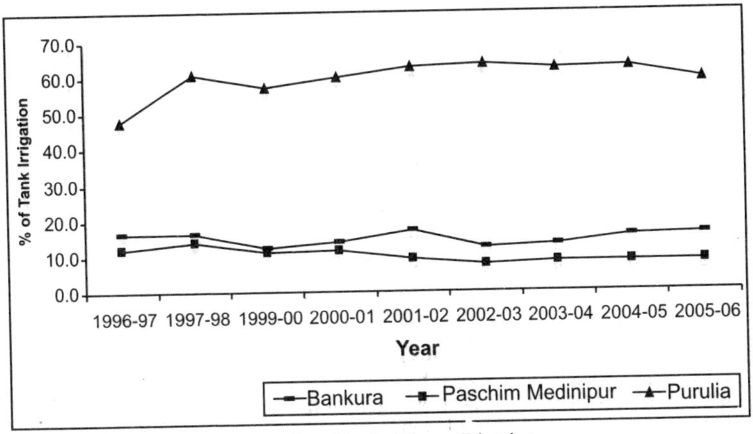

Fig. 2.5: Per cent of Tank Irrigation in Three Districts

2.10 Block-wise Tank Irrigation Development in the Districts under Consideration

2.10.1 Tank Irrigation in Bankura

Total irrigated area from tanks in the district of Bankura has decreased from 50.6 thousand hectares in 1993-94 to 33.1 thousand hectares in 2005-06. The block-wise 4-year average for two sub-periods 2001-05 and 1996-2000 have been calculated and presented in the following Table. As Table 2.11 shows out of 21 blocks in Bankura district, tank irrigation has decreased in 18 blocks. The blocks which have experienced more decrease in tank irrigation are Chhatna, Ranibandh, Indpur etc. Table 2.11 also presents that the share of tank irrigated

Table 2.11: Block-wise Average Annual Tank Irrigated in Bankura District

Blocks	*Tank Irrigated area ('00 hec)*			*% of Tank Irrigated Area*		
	2001-2005	*1996-2000*	*Change in tank irrigated area*	*2001-2005*	*1996-2000*	*% Change in tank irrigated area*
1. Bankura-I	1817.6	1081.0	736.6	65.1	30.7	34.4
2. Bankura-II	1094	1930.6	–836.6	40.8	27.9	12.8
3. Chhatna	1231.2	2298.4	–1067.2	72.4	71.5	0.8
4. Saltora	844.2	1543.2	–699.0	82.3	79.9	2.4
5. Mejhia	1514.4	1507.0	7.4	54.2	32.2	22.1
6. G. Ghati	7214	7880.0	–666.0	82.4	77.4	5.0
7. Barjora	990	1422.0	–432.0	10.0	12.5	–2.5
8. Onda	2787.6	2301.0	486.6	16.3	8.8	7.5
9. Indpur	1972	3940.0	–1968.0	29.0	39.7	–10.8
10. Khatra (Khatra-1)	1470	1740.0	–270.0	20.1	21.2	–1.1
11. Hirbandh (Khatra-2)	587.2	1294.2	–707.0	24.3	40.4	–16.1
12. Ranibandh	5620	6869.0	–1249.0	74.1	74.0	0.2
13. Taldangra	988.6	716.0	272.6	8.6	5.8	2.8
14. Simlapal	1850	2310.4	–460.4	9.3	9.5	–0.2
15. Raipur (Raipur-1)	2548	2320.0	228.0	11.8	10.8	1.0
16. Sarenga (Raipur-2)	170	633.0	–463.0	1.0	3.2	–2.1
17. Bishnupur	296	366.0	–70.0	1.7	1.5	0.2
18. Joypur	2789	2862.2	–73.2	12.8	9.8	3.0
19. Kotulpur	230	2076.0	–1846.0	1.3	7.5	–6.2
20. Sonamukhi	302.4	650.6	–348.2	1.5	2.4	–0.9
21. Patrasayer	142	303.0	–161.0	0.8	1.6	–0.8
22. Indus	353	394.0	–41.0	1.7	1.7	0.0
Total	36811	46437	9626	28.2	25.9	2.3

Source: Government of West Bengal, District Statistical Handbook, Bankura, Various Issues.

area in total irrigated area in different blocks for the two sub-periods. In percentage terms tank irrigated area have decreased in Indpur, Hirbandh and Kotulpur.

Percentage of tank irrigated area for the different blocks is presented in the following Figure on he basis of Minor Irrigation Census. The higher dependence on tank irrigation is seen in the blocks of Khatra-1, Chhatna, Bankura-1, Rani Bund etc.

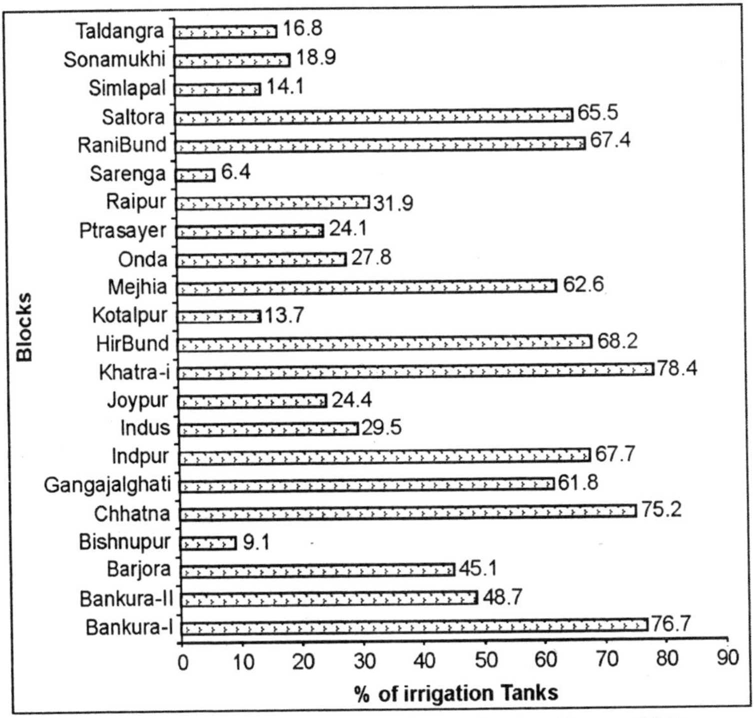

Source: Government of West Bengal, Minor Irrigation Census, 2001.

Fig. 2.6: Percentages of Tank Used for Irrigation Purpose in Bankura District

2.10.2 Block-wise Tank Irrigation in Paschim Medinipur

Block-wise tank irrigation development is presented in Table 2.12. We have compared here average annual tank irrigation area for the two sub-periods 1996-2000 and 2001-06. Out of 29 blocks we see that tank irrigated area has fallen in 20 blocks. Tank irrigated area has decreased more in Salboni, Keshpur, Garbeta-3 etc. Table 2.12

also presents that the share of tank irrigated area in total irrigated area in different blocks for the two sub-periods. In percentage terms tank irrigated area has decreased in Nayagram, Sankrail, Garhbeta-3.

Table 2.12: Block-wise Average Annual Tank Irrigated in Pachim Medinipur District

	Blocks	Tank Irrigated area ('00 hec)			% of Tank Irrigated Area		
		2001-06	1996-2000	Change	2001-06	1996-2000	Change
1	Jhargram	1900	1244.5	655.5	12.7	12.5	0.2
2	Binpur-1	195	309.5	–114.5	0.9	2.1	–1.2
3	Binpur-2	435	590	–155	5.2	8.8	–3.6
4	Jamboni	1151	790	361	11.5	16.1	–4.6
5	Nayagram	188	566	–378	2.9	19.2	–16.3
6	Sankrail	180	632.5	–452.5	2.4	7.5	–5.2
7	Gopiballavpur-1	784.8	995	–210.2	6.9	14.7	–7.8
8	Gopiballavpur-2	302	588.75	–286.75	2.9	13.1	–10.1
9	Salboni	544	1655	–1111	9.2	16.8	–7.6
10	Keshpur	730	2295	–1565	7.9	12.4	–4.4
11	Garbeta-1	183.8	601	–417.2	1.1	4.6	–3.5
12	Garbeta-2	162	347.5	–185.5	1.6	3.2	–1.6
13	Garbeta-3	115.2	893	–777.8	1.9	10.9	–9.0
14	Midnapur	580	475	105	8.9	10.1	–1.2
15	Debra	1666	445	1221	8.9	2.9	6.0
16	Pinbla	2000	1535	465	14.4	11.7	2.7
17	Keshiary	300	397.5	–97.5	5.5	5.8	–0.4
18	Datan-1	410	210	200	4.3	3.8	0.6
19	Datan-2	55	245	–190	1.1	4.4	–3.3
20	Narayangarh	560	895	–335	5.5	7.9	–2.4
21	Mohonpur	2000	875	1125	28.4	15.7	12.7
22	Sabaong	3000	4206	–1206	18.0	25.6	–7.6
23	Khargapur-1	600	575	25	12.2	15.3	–3.1
24	Khargapur-2	40	142.75	–102.75	0.3	1.3	–1.0
25	Chandrakana-1	600	2300	–1700	6.6	26.1	–19.5
26	Chandrakana-2	400	2000	–1600	7.7	25.9	–18.2
27	Ghatal	3000	1900	1100	22.1	16.6	5.6
28	Daspur-1	930	2200	–1270	9.4	23.3	–14.0
29	Daspur-2	1500	2100	-600	17.0	23.0	–6.0
	Total	24511.8	32009.0	7497.2	8.2	12.5	–4.3

Source: Government of West Bengal, District Statistical Handbook, Paschim Medinipur, Various Issues.

Percentage of tanks used for irrigation purposes in different blocks of Paschim Medinipur is presented in the following Figure. The blocks with higher percentages of tanks used for irrigation purpose are given in Fig. 2.7.

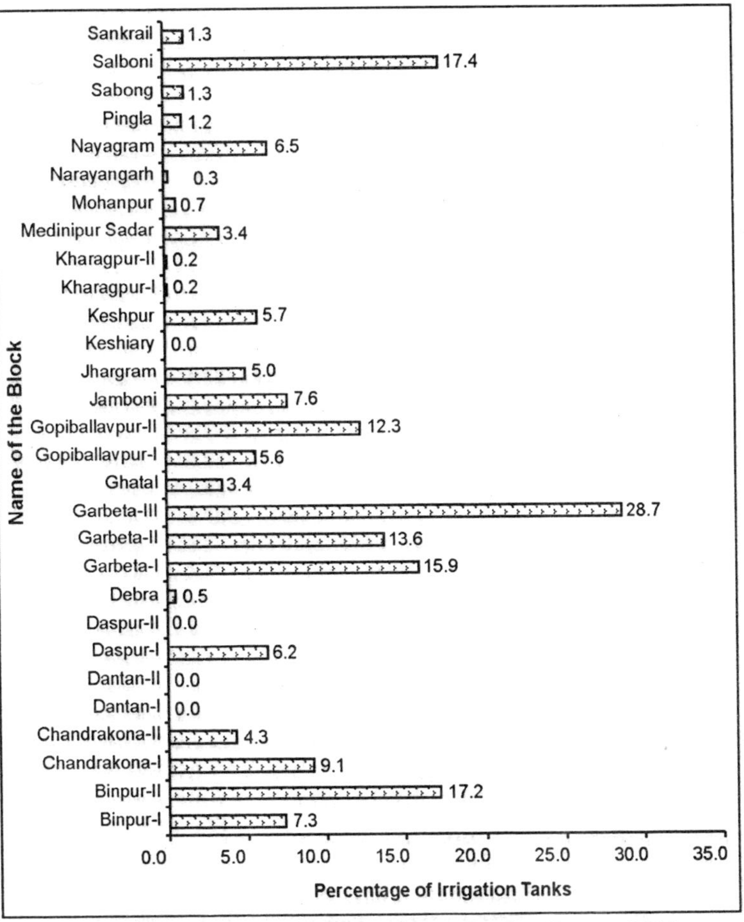

Source: Government of West Bengal, Minor Irrigation Census, 2001.

Fig. 2.7: Percentage of Tank Used for Irrigation Purpose in Paschim Medinipur District (2001)

2.10.3 Tank Irrigation Development in Purulia

Block-wise tank irrigation development is presented in the following Table 2.13. We have compared here average annual tank irrigation area for the two sub-periods 1996-2000 and 2001-06. Out of 20 blocks in Purulia we see that tank irrigated area has fallen in 14 blocks. Tank irrigated area has decreased more in the blocks of Hura, Jaypur, Baghmundi etc. Table 2.14 also presents that the share of tank irrigated area in total irrigated area in different blocks for the two sub-periods. In percentage terms tank irrigated area have decreased in Raghunathpur-II, Jaypur, Kashipur etc.

Table 2.13: Block-wise Tank Irrigation in Purulia District

Sl. No.	Blocks	Tank Irrigated Area ('00 hec)			% of Tank Irrigated Area		
		2001-2005	1996-2000	Change	2001-2005	1996-2000	Change
1.	Arsha	480.6	876.4	–395.8	7.9	19.0	–11.1
2.	Baghmundi	348.4	1111.6	–763.2	14.4	25.2	–10.8
3.	Balarampur	973.4	1411.6	–438.2	34.0	36.4	–2.4
4.	Barabazar	1055.2	1990.4	–935.2	22.1	30.8	–8.7
5.	Jaypur	1318	2466.2	–1148.2	30.8	70.5	–39.7
6.	Jhalda-I	1495.4	1135	360.4	41.8	23.7	18.1
7.	Jhalda-II	3659.8	1828	1831.8	57.6	38.9	18.6
8.	Bandowan	1980.6	2611	–630.4	48.7	59.6	–10.9
9.	Hura	2027	3233.4	–1206.4	58.5	78.2	–19.7
10.	Manbazar-I	1152.4	2187.4	–1035	69.6	65.0	4.6
11.	Manbazar-II	991.8	1565.8	–574	45.1	61.5	–16.5
12.	Puncha	1677	1728.6	–51.6	73.7	68.7	5.0
13.	Purulia-I	968	1159.2	–191.2	44.9	48.4	–3.5
14.	Purulia-II	1141.6	1632.4	–490.8	52.0	65.3	–13.2
15.	Kashipur	227.4	830.4	–603	9.6	29.4	–19.8
16.	Neturia	901.2	670.4	230.8	77.3	46.3	31.1
17.	Para	2669.8	2130.6	539.2	51.7	51.9	–0.1
18.	Raghunathpur-I	1281	984.2	296.8	68.5	64.6	3.9
19.	Raghunathpur-II	139.8	908.8	–769	12.0	64.2	–52.2
20.	Santuri	1201.8	855.4	346.4	33.5	42.9	–9.4
	Total	25690	31317	5627	42.7	49.5	–6.8

Source: Government of West Bengal, District Statistical Handbook, Purulia, Various Issues.

Percentage of tanks used for irrigation purposes in different blocks of Purulia is presented in the following Figure. Tanks are widely used for irrigation purpose in the district. In Joypur, Santuri, Purulia-II, Barabazar more than 80 per cent tanks are used for irrigation purpose.

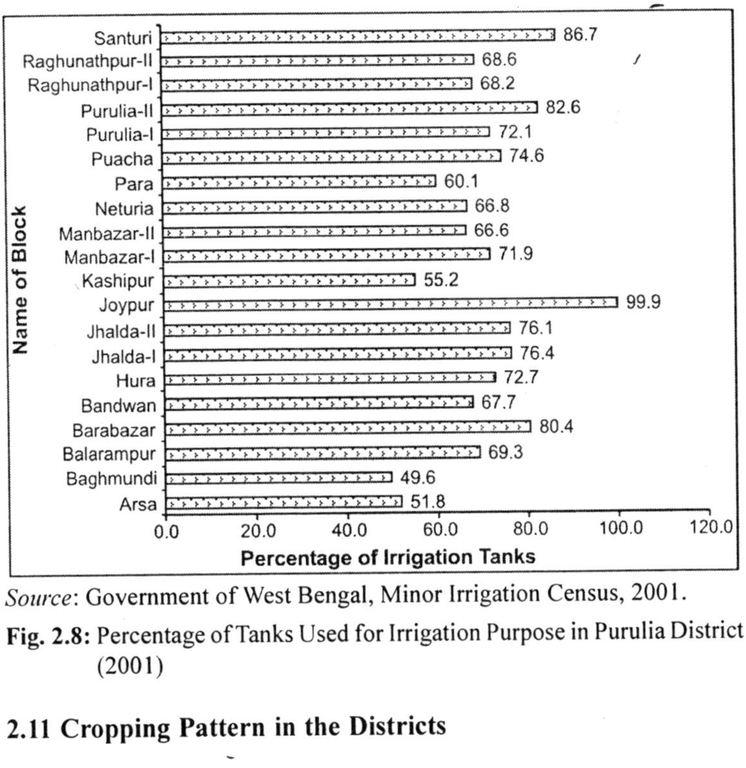

Source: Government of West Bengal, Minor Irrigation Census, 2001.

Fig. 2.8: Percentage of Tanks Used for Irrigation Purpose in Purulia District (2001)

2.11 Cropping Pattern in the Districts

We see that the Aman paddy is the most dominant crop in all the districts, though the percentage of area for aman over the years has decreased to some extent and that the percentage of boro has increased. Crops grown in the district are paddy (aus, aman and boro); wheat, potato, vegetables, oilseeds, pulses, sugarcane, jute etc. Major crop-wise percentage area is shown in Table 2.14.

2.11.1 Percentage of Boro Paddy to Total Rice in the Three Districts

Percentage of boro area as a percentage of total paddy area is an indicator of irrigation requirement as boro paddy is water intensive crop (crop-wise water requirement is given in the Appendix of Chapter 1). As the figure shows percentage of boro cultivation in Paschim Medinipur has increased at rapid rate compared to other districts. In fact the percentage of boro crop is very low in Purulia which probably shows the irrigation shortage in the district.

Table 2.14: Percentages Area of Different Crops (2006)

Crops	Area ('00 hec)			% area		
	Paschim Medinipur	Bankura	Purulia	Paschim Medinipur	Bankura	Purulia
Aus	66.358	19.366	0.780	5.30	4.65	0.24
Aman	745.200	287.815	287.528	59.46	69.09	88.44
Boro	259.944	39.644	1.354	20.74	9.52	0.42
Wheat	12.000	8.600	17.000	0.96	2.06	5.23
Maize	8.000	0.700	11.200	0.64	0.17	3.44
Other Cereals	0.000	0.350	0.600	0.00	0.08	0.18
Total Pulses	0.400	0.150	0.450	0.03	0.04	0.14
Mustard	25.700	12.200	1.300	2.05	2.93	0.40
Linseed	0.050	0.050	0.200	0.00	0.01	0.06
Til	54.500	19.500	0.800	4.35	4.68	0.25
Fibre Crops	7.050	0.550	0.300	0.56	0.13	0.09
Sugarcane	2.200	0.050	0.700	0.18	0.01	0.22
Tobaco	0.000	0.100	0.000	0.00	0.02	0.00
Potato	60.200	25.900	1.600	4.80	6.22	0.49
Dry Chilli	9.100	1.400	1.200	0.73	0.34	0.37
Ginger	2.500	0.200	0.100	0.20	0.05	0.03
	1253.202	416.575	325.112	100.00	100.00	100.00

Source: Based on District Staitistical Handbook, Government of West Bengal.

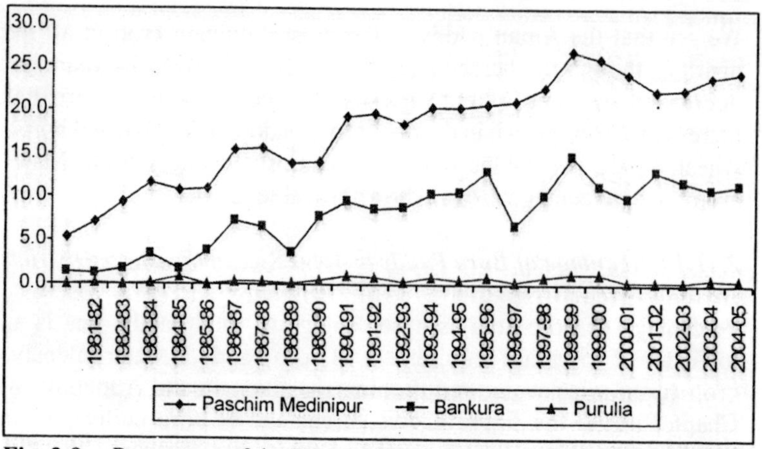

Fig. 2.9: Percentage of the Boro Paddy to Total Rice in the three Districts

2.12 Relation between Rainfall and Tank Irrigation in the Districts

Most of the tanks in the districts are rain-fed. Obviously, rainfall has an effect on the performance of tank irrigation and on paddy yields. We have collected the month-wise data on rainfall for all the districts. Around 80 per cent of the rainfall in the districts is received during the month of June to September. It may be mentioned that the rainfall during the period of June-July is critical for the transplantation of the Aman crop. Rainfall and tank irrigated area are closely associated variables. Higher rainfall is expected to have positive impact on tank-irrigated area. To examine the relationship we have prepared the two columns for all the districts—index of tank irrigated area and percentage deviation of the rainfall for the period (June to September) from the normal rainfall. It may be noted that the rainfall in the districts have slightly lower tendency and deviations are high in the latter years and the tank-irrigated area electived heavily during the latter years.

Table 2.15: Relationship between Areas Irrigated by Tank and Rainfall during 1993 to 2005 in Bankura, Paschim Medinipur, Purulia District

Years	*Bankura*		*Paschim Medinipur*		*Purulia*	
	Index of Tank Irrigated Area	*Deviation of Rainfall from Normal (June to September)*	*Index of Tank Irrigated Area*	*Deviation of Rainfall from Normal (June to September)*	*Index of Tank Irrigated Area*	*Deviation of Rainfall from Normal (June to September)*
1993	100.00	16.50	—	—	—	—
1994	82.41	7.67	—	—	100.00	16.48
1995	90.91	1.01	100.00	2.19	63.12	17.88
1996	99.60	10.94	93.19	3.32	66.03	20.79
1997	104.74	2.11	117.01	22.65	33.53	9.43
1998	88.34	–37.71	100.55	–57.27	45.92	–30.72
1999	88.14	33.44	85.43	16.98	42.13	14.69
2000	78.06	–10.95	84.04	2.80	40.52	–31.04
2001	70.95	–8.89	76.44	3.49	32.65	6.59
2002	80.04	6.94	56.67	0.57	33.24	4.94
2003	83.40	–22.67	63.48	–29.84	39.50	–27.16
2004	64.03	–15.31	62.92	–1.56	40.52	–7.00
2005	65.42	–31.70	61.74	–1.56	41.25	–48.24

The line diagram showing the relationship between the two variables, namely rainfall variation and tank irrigation for Bankura is presented in Fig. 2.10.

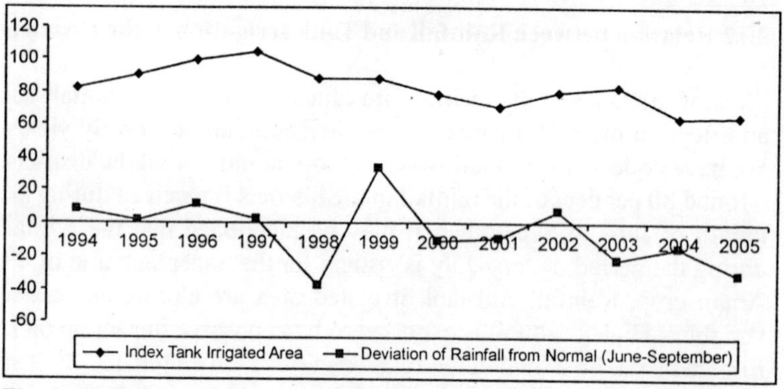

Fig. 2.10: Relationship between Tank Irrigated Area and Rainfall during 1993 to 2005: Bankura District

The line diagram showing the relationship between the two variables is presented in Figure 2.11.

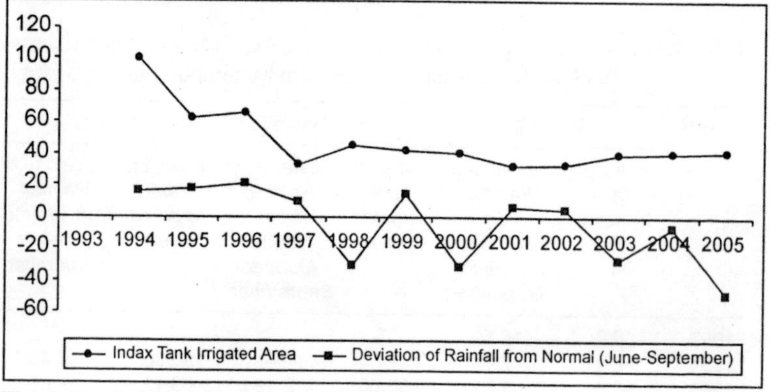

Fig. 2.11: Relationship between Areas Irrigated by Tank and Rainfall during 1994 to 2005 in Purulia District

The line diagram showing the relationship between the two variables for Paschim Medinipur district is presented in Fig. 2.12.

2.13 NREGS in Tank Irrigation

NREGA is a wage employment programme implemented in the country under National Rural Employment Guarantee Act (NREGA) 2005, notified on 7th September, was brought into force by the Union Government in February 2006 with twin objectives in mind. It ensured

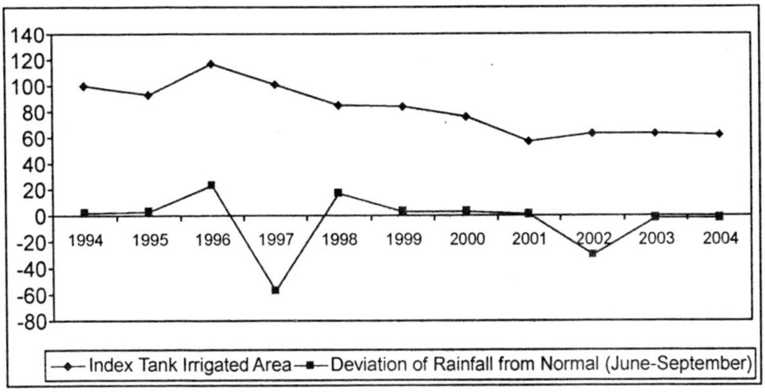

Fig. 2.12: Relationship between Areas Irrigated by Tank and Rainfall during 1995 to 2005 in Paschim Medinipur District

the legal right to work for a hundred days to poor people whoever is willing to work at a minimum wage rate, particularly in the rural areas, which in turn would reduce the flow of rural to urban migration. The programme has opened up a great scope for rural infrastructure development including water conservation and water harvesting. In the Table below we have presented district-wise expenditure on different components. The district-wise expenditure for water conservation and water harvesting are as follows: Bankura – 2822 lakh, Paschim Medinipur – 3111 lakh, Purulia – 1697 lakh.

Table 2.16: Works under MGNREGA during the Financial Year (Rs. Lakh) 2010-2011

	Bankura	*Paschim Medinipur*	*Purulia*
Rural Connectivity	2231.88	2788.14	649.80
Flood Control	199.06	272.12	1.36
Water Conservation and Water Harvesting	2822.31	3111.23	1697.90
Drought Proofing	88.83	107.46	46.86
Micro Irrigation	357.20	326.10	4.34
Provision of Irrigation Facility to Land Development	49.53	814.21	106.37
Renovation of Traditional Water Bodies	2075.77	2927.25	46.07
Land Development	274.65	652.46	132.97
Any Other Activity Approved by MRD	22.09	29.50	13.00
Total	**8121.31**	**11028.47**	**2698.67**

Source: www.nrega.nic.in

In the Table below we have presented the percentage expenditure on different schemes. As we see the percentage expenditure on water conservation and water harvesting are as follows: Bankura – 34.8 per cent, Paschim Medinipur – 28.2 per cent and Purulia 62.9 per cent implying Purulia's importance on this scheme.

Table 2.17: Works under MGNREGA during the Financial Year 2010-2011 (%)

	Expenditure (%)		
	Bankura	*Paschim Medinipur*	*Purulia*
Rural Connectivity	27.5	25.3	24.1
Flood Control	2.5	2.5	0.1
Water Conservation and Water Harvesting	34.8	28.2	62.9
Drought Proofing	1.1	1.0	1.7
Micro Irrigation	4.4	3.0	0.2
Provision of Irrigation Facility to Land Development	0.6	7.4	3.9
Renovation of Traditional Water Bodies	25.6	26.5	1.7
Land Development	3.4	5.9	4.9
Any Other Activity Approved by MRD	0.3	0.3	0.5
Total	100.0	100.0	100.0

Source: Calculation based on www.nrega.nic.in

Summary of the Chapter

In this chapter we have mainly dealt with agro-climatic characteristics of the sample districts, namely Bankura, Purulia and Paschim Medinipur. The analysis has been done in comparative framework. The rainfall pattern shows that the average rainfall for the period 1980-2005 in the districts has similarity with wider fluctuations in Paschim Medinipur in the latter years. The source-wise irrigation development for all the three districts for the period 1996-97 to 2005-06 has been given. Block-wise tank irrigated area in the three districts and their changes for the two sub-periods (1996-2000) and (2001-2005) have been compared. The results reveal that in many blocks the tank irrigated area has declined both in absolute and percentage terms. We have also investigated the relation between tank irrigated area and rainfall in the districts. NREGS expenditure pattern shows Purulia gives more emphasis on water harvesting compared to the other two districts.

SELECTED PICTURES OF TANKS/BUNDS UNDER STUDY

Plate 1: Kechanda *Raj Bund*, Kechanda, Paschim. Medinipur

Plate 2: *Banuli Bund*, Chhatna, Bankura

"HAPPA" (5% MODEL)

Plate 3: Hirbandh, Bankura

"INDARA"

Plate 4: Berada, Barabazar, Purulia

CHECK DAM

Plate 5: Bankura-1, Bankura

JORH

Plate 6: Berada, Barabazar, Purulia

IRRIGATION SYSTEM

Plate 7: Irrigation by Dobka, Purulia

USE OF TANK - FISHING

Plate 8: Gopiballavpur-1, Paschim.Medinipur

USE OF TANK - BATHING

Plate 9: Susunia, Chhatna-1, Bankura

IRRIGATED LAND FROM TANKS

Plate 10: Barabazar, Purulia

PROBLEMS OF BUND

Plate 11: Supply Channel Damage, *Ghosh Bund*, Paschim.Medinipur

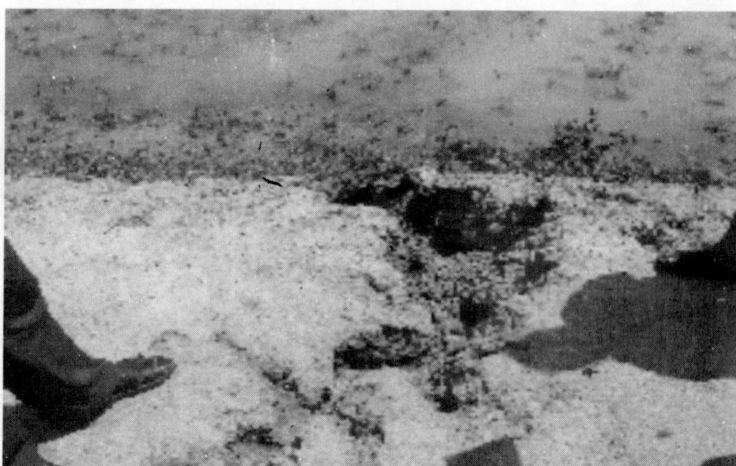

Plate 12: Breached Tank Bund, Lodhasuli, Paschim Medinipur

Plate 13: Damage of Supply Channel, Chhatna, Bankura

Plate 14: Low Height of Tank Embankment, Hura, Purulia

Plate 15: Lock Gate Damage, *Mela Bund*, Paschim Medinipur

Plate 16: Cultivation in Tank-bed, Saltoria Bund, Paschim Medinipur

Chapter 3

Tank Study: Primary Survey

In the previous chapter, i.e. Chapter 2, we discussed about various features relating to tank irrigation for selected three districts of our study, namely Bankura, Paschim Medinipur and Purulia. We have already stated that we have selected 181 tanks and 533 households for our study from the selected districts. In this chapter, we will present the results relating to different features of tank irrigation for the selected tanks under study. First, we shall give the names of the tanks of our study and then we shall present the different characteristics of the tanks. In the end we shall analyse the efficiency of the tanks.

3.1 Tanks Surveyed under Study Villages

We have selected villages which lie in the dry zones of the districts which are mostly in the western parts of the districts. In the regions of Western Bankura, Western Paschim Medinipur and Purulia the dominant soil type is red and brown soil. Steep uplands are eroded because of chemical weathering and sediments are accumulated in valleys. The upland soils are reddish and the topland soils are brownish because of greater hydration of iron oxides in the lower region. Soils are sandy loams with concretions. The porous nature of the soil allows rapid penetrations of water along with nutrients. This is more so in the higher tracts. The soils are thus poor in organic matter. The range of soil *p*H is between 6 and 6.5 making it acidic in nature. The brown soils are more fertile than red soils. In these types of soils inputs of more irrigation and fertilizer are required. Soil texture is the most important feature of soil from the point of view of irrigation. More irrigation is needed in case of high permeability. In these regions high sand contents make the soil more permeable. Soil acidity/alkalinity is also important as crops are diversely responsive to this character. Irrigation water requirement also depends upon the *p*H. So, this zone consisting of the sub-humid hilly to undulating plateau and plateau and plateau fringe tracts possess distinctive problems of its own. The inherently sandy structured soil of this region favours rapid percolation of the water to deeper layers. Soils remain dry necessitating irrigation. The undulating terrain and sandy soil pose a problem for these regions (*Chatterjee 1995*).

3.1.1 Tanks Surveyed in Bankura

As mentioned earliar we have surveyed 181 tanks from 102 villages scattered over 3 districts. In Bankura 52 tanks have been surveyed from 26 villages covering 5 GPs, namely Chhatna-1, Chhatna-2, Susunia, Jagdalla-2 and Kalpathar. Highest village area is at Jaljalia village and the highest population is at Kantasole village. Also from our survey we see that the highest number of households are at Bamunkali village and the highest family size is at Hatisal. The information is collected from primary field survey. From our survey we see that the highest total agricultural land is Kalaiberia village (Chhatna 2 block) and the highest irrigated land is Baidka village (Jagadalla-2) followed by Jamadarara village. The averages for different parameters for a village out of 26 villages in sample district Bankura are as follows: Area-1.4 sq. km., total projected population (2008)-692.5, number of households (2001)-120.9, population density per sq. km.-517.6, family size-5.5, Agricultural land-264.6 acre, Irrigated Land-126.2 acre, Numbers of Irrigation Tanks-3.7, Numbers of Other Irrigation Sources-0.5, Electricity-1 (100%), Distance of village from Paka Road-2 km.

In Table 3.1, we have presented the surveyed tanks of our study with corresponding GPs and Blocks to which tanks belong.

3.1.2 Tanks Surveyed in Paschim Medinipur

Here we show the description of the Villages in Paschim Medinipur district. The information is collected from secondary sources. From our survey we see that the highest village area is at Dubra village. The highest population and the highest number of households are at Belia village. From our survey we see that the highest total agricultural land

Table 3.1: Name of the Tanks Surveyed in Bankura

Sl. No.	*Name of Tanks*	*Ownership*	*Area (acre)*	*Village*	*GP*	*Block*
1	Bangram Natun Pukur	Community	2.6	Bangram	Chatna-2	Chatna
2	Madhusudan Bund	Community	6.6	Kalaiberia		Chatna
3	Banka Bund	Community	30.0	Gurhputa		
4	Upar Bund	Community	6.0			
5	Choto Pukur (Bamun Bund)	Community	8.0	Kuludi		
6	Kanak Sayer	Community	5.0	Ghoramuli		
7	Dighi	Community	5.0			
8	Ghoramuli Barha Bund	Community	16.5			
9	Khana Pukur	Community	4.0	Hatisal		
10	Jhagrapur Barha Bund	Community	22.0	Jhagrapur		
11	Raypukur	Personal	3.3	Bamunkali		
12	Khirki	Personal	5.2			
13	Babur Bund	Personal	4.5	Mandaspara		
14	LaBunda Bund	Personal	5.0	Susunia	Susunia	
15	Upar Bund	Personal	29.7	Bharatpur	Susunia	
16	Tantir Bund	Personal	6.6			
17	Ghosal Bund	Personal	5.9	Bansole and Jaljalia		
18	Sikdargorhia	Personal	4.0	Jaljalia		
19	Bamunigerhia	Personal	1.0			
20	Mathura Bund	Personal	2.0			
21	Bishurgerhia	Personal	1.3			
22	Namo Bund	Personal	3.3	Bansole		
23	Bankrhageria	Personal	1.3			
24	Kishoregeria	Personal	1.3			
25	Nekrhegeria	Personal	2.0			
26	Upar Bund	Personal	8.3	Gidhudia		
27	Goyal Bund	Personal	2.3			
28	Arhsata	Personal	6.6			
29	Basuli Bund	Personal	10.0	Kadra	Chatna-1	
30	Saheb Bund	Government	8.0			
31	Ranga Pukur	Personal	5.0	Hari Bundi		
32	Kadma Pukur	Personal	2.0			
33	Bara Bund	Community	10.0			
34	Majhi Pukur	Community	4.0			
35	Natun Pukur	Personal	3.2	Kantasole		
36	Padma Pukur	Personal	4.0	Seuligeria		
37	Ranga Pukur	Personal	3.8			
38	Purano Pukur	Personal	3.0	Sarberhia		
39	Natun Pukur	Personal	3.0			
40	Goyala Bund	Personal	3.0			
41	Kulur Bund	Personal	8.9	Jamadarara		
42	Kunrher Bnadh	Community	1.3	Sonamelia	Jagadalla-2	Bankura-1
43	Bamun Bund	Community	3.3			
44	Mansar Bund	Community	1.7			
45	Gopal Sayer	Personal	4.0			
46	Barha Bund	Community	23.1	Baidka		
47	Natun Bund	Community	16.5			
48	Chakalta	Community	6.6			
49	Bhaturi Bara Pukur	Community	19.7	Bhaturi		
50	Dubraji Bara Bund	Community	3.6	Dubraji		
51	Purna Baurir Hapa	Personal	0.1	Nekrhagorhiya	Kalpathar	
52	Sudhir Maler Hapa	Personal	0.3	Pathrakata		

Source: Primary (Field) Survey.

is in Jarulia and the high percentage of irrigated lands is observed in Kendugari village, Shyamnagar village and Kechanda village. Also seen that the highest numbers of Tanks are in Hijragh village and the highest numbers of other sources of irrigation are in Belia village. Moreover, electricity is not available in Saltoria village and Banskona village. The averages for different parameters of the villages are as follows: Area–1.4 sq. km., total projected population (2008)–584.0, number of households (2001)–104.3, population density per sq. km.–410.6, family size–5.1, Agricultural land–224.6 acre, Irrigated Land–106.0 acre, numbers of Irrigation Tanks–2.2, numbers of Other Irrigation Sources–1.8, Electricity–1 (100%), Distance village from Paka Road– 3.1 km.

Table 3.2 shows names of tanks in Paschim Medinipur with corresponding GPs and Blocks. As given in Table 3.2, we have surveyed 69 tanks in the Paschim Medinipur district covering 13 GPs and 6 blocks. The names with their areas are given in Table 3.2. The bund with highest area is observed to be Kechanda in Paschim Medinipur.

3.1.3 Tanks Surveyed in Purulia

Here we present the description of the Villages under Purulia district. The information is collected from the primary survey source. From our survey we see that the highest village area is at Keshargarh village and the highest population and the highest numbers of households are at Para village. The information is collected from primary field survey. From our survey we see that the highest agricultural land is in Para village and the highest percentage of irrigated land is in Asanmoni village. The highest numbers of irrigated tanks is in Sankra village and other sources of irrigation are available only in Madandi village. Electricity is available in every village under study in Purulia district. The averages for different parameters per village are as follows: Area–4.2 sq. km., total population (2008)–2236, number of households (2001)–389.7, population density per sq. km.–600.5, family size-6.1, Agricultural land–448.5 acre, Irrigated Land–186.3 acre, numbers of Irrigation Tanks–6.1, numbers of Other Irrigation Sources–0.04, Electricity–100 per cent, Distance village from Paka Road-1.3 km.

The surveyed tank names and their locations in Purulia district are shown in Table 3.3. From Table 3.3, we see that our surveyed

Table 3.2: Name of the Tanks Surveyed in Paschim Medinipur

Sl. No.	*Name of Tanks*	*Ownership*	*Area (acre)*	*Village*	*GP*	*Block*
1	Bankar Bund	Personal	6.0	Bandar Bhula	Jhargram (muni.)	Jhargram
2	Ray Pukur-1	Personal	3.2	Pukuria	Sapdhara	
3	Ray Pukur-2	Personal	2.3			
4	Ray Pukur-3	Personal	3.0			
5	Ganyer Bund	Govt.	25.2	Gajasimul	Lodhasuli	
6	Patharchakri Khash Bund	Govt.	2.9	Patharchakri		
7	Saltoria Bund	Personal	21.0	Saltoria		
8	Saontaldiha Bund	Govt.	8.4	Saontaldiha		
9	Saontaldiha Desh Bund	Govt.	4.2	Shyamnagar		
10	Dahatmul Padma Pukur	Community	21.8	Dahatmul (Daskhin)		
11	Dahatmul Sech Pukur	Personal	0.8			
12	Ghosh Bund	Personal	63.0	Labkush	Salbani-6	
13	Mehera Bund	Govt.	12.4	Barhabanpura		
14	Banpura Desh Bund	Community	6.3	Bamanmara		
15	Mahato Bund	Personal	5.0	Salgerhia		
16	Tentul Bund	Govt.	3.4	BaladhBund		
17	Ela Bund	Personal	4.2	Baghmurhi		
18	Padra Bund	Community	5.0	KamarBundi		
19	Rilif Bund	Community	11.1	Jarulia	Nedhabaharha-5	
20	Mahato Bund	Personal	5.0	Chakua		
21	Auligeria Sech Bund	Govt.	6.2	Auligerha		
22	Kechanda Raj Bund	Personal	200.0	Kechanda	Radhanagar-1	
23	Mela Bund	Govt.	27.3	Radhanagar		

(*Contd.*)

Table 3.2 (*Contd.*)

Sl. No.	*Name of Tanks*	*Ownership*	*Area (acre)*	*Village*	*GP*	*Block*
24	Belia Marha Pukur	Personal	9.2	Belia	Dubrha-7	Jamboni
25	Kedukota Bund	Personal	2.8			
26	Sabarparh Bund	Govt.	2.8			
30	Gopalpur Sabar Bund	Personal	1.3	Gopalpur		
31	Jamui Pukur	Govt.	0.8	Jamui		
32	Palbansi Sech Pukur	Govt.	0.8	Palbansi		
33	Palbansi Rajar Bund	Govt.	2.9			
34	Padma Pukur	Govt.	1.5	Dubra		
35	Batabani Pukur	Govt.	1.5			
36	Kurharia Bund	Govt.	10.5	Kurharia		
27	Baghuya Pukur	Govt.	2.1	Baghuya	Jamboni-8	
28	Choto Ghong Pukur	Govt.	1.7	Choto Ghong		
29	Maiti Pukur	Govt.	2.5	Harinagang		
37	Dighı	Personal	9.6	Belar	Lalua	Keshiary
38	Pana Pukur	Personal	1.0			
39	Ranader Pukur (East)	Personal	1.5			
40	Ranader Pukur (Near Road)	Personal	2.0			
41	Guri Pukur	Community	3.0	Fandar		
42	Sansan Pukur	Community	1.5			
43	Fandar High School Pukur	Community	0.7			
44	Chakulia	Govt.	6.0	Panchiar		
45	Raygeria	Personal	3.0			
46	Cherua Pukur	Personal	3.0			
47	Bat-tolia Pukur	Personal	2.5	Hijragh		

(*Contd.*)

Table 3.2 (*Contd.*)

Sl. No.	*Name of Tanks*	*Ownership*	*Area (acre)*	*Village*	*GP*	*Block*
48	Pon Bund	Community	2.5	Kashijorha	Kashijorha-9	Salbani
49	Mana Bund	Personal	6.3			
50	Kashijorha Barha Bund	Personal	4.2			
51	Mahisharha Bund	Govt.	7.1	Mahisharha		
52	Janakpur Bund	Govt.	0.6	Janakpur		
53	Barju Pukur	Govt.	0.6	Barju		
54	Simul Bund	Govt.	7.7	Banskopna		
55	Banskona Bamun Bund	Personal	6.2	Banskona		
56	Bara Bund	Community	7.6	Pindrakuli		
57	Pindrakuli Choto Pukur	Personal	1.7			
58	Baragada Hadahadi Pukur	Govt.	1.9	Baragadha		
59	Kharikasuli Pukur	Personal	0.9	Kharikasuli	Chandra-2	Medinipore Sadar
60	Haniyachatri Bund	Govt.	7.7	Dainmari		
61	Natun Pukur	Govt.	0.4	Kharikasuli		
62	Rabindranath Mandir Pukur	Personal	0.3	Sarengashole	Manidaha-3	
63	Chuturam Mandir Pukur	Personal	0.2			
64	Pindrasole Pukur	Personal	0.5	Pindrasole		
65	Dhansole RSVY Pukur	Personal	0.3	Dhansole	Kendugari-7	Gopiballavpur-1
66	Chitamatia Gramer Pukur	Community	0.3	Chitamatia		
67	Kendugari Band	Community	7.4	Kendugari		
68	Lohamaliya Pukur	Personal	1.6	Lohamaliya	Gopiballavpur-5	
69	Chowdhuri Pukur	Personal	1.3			

Source: Primary (Field) Survey.

Table 3.3: Name of the Tanks Surveyed in Purulia

Sl. No.	*Name of Tank*	*Ownership*	*Area (acre)*	*Village*	*GP*	*Block*
1	Nim Bund	Community	4.7	Nimdiha	Lodhurka	Hura
2	Doli Bund	Community	4.0		G.P.	
3	Shayar Bund	Community	2.7			
4	Santiran Bund	Personal	3.3			
5	Bara Bund	Community	6.0	Chitra		
6	Bener Bund	Community	4.0			
7	Jola Bund	Community	3.0			
8	Hullung Bara Bund	Community	8.3	Hullung	Jabardra	
9	Mahato Bund	Community	3.0	Kulabahal	G.P.	
10	Singher Bund	Community	2.7			
11	Sunri Bund	Community	3.0			
12	Sizu- Makhana Bund	Community	16.0	Sizu-Makha	Keshargarh	
13	Kasi Bund	Community	10.0	Keshargarh	G.P.	
14	Kalia Bund	Govt.	12.0			
15	Bighasaheb Bund	Community	1.2	Arjunjora		
16	Sati Bund	Community	0.5			
17	Jojodi Bara Bund	Govt.	11.3	Jojodi		
18	Jojodi Purano Bund	Community	4.0			
19	Fulgeria Bund	Personal	1.0			
20	Bucha Bund	Community	2.0	Madandi		
21	Kengeria Bund	Community	2.7			
22	Jahugeria Bund	Community	2.0			
23	Head Bund	Community	6.0	Chirumarcha	Hutmura	Purulia-2
24	Birgora Bund	Govt.	3.3			
25	Mahato Bund	Community	3.3			

(Contd.)

Table 3.3 (*Contd.*)

Sl. No.	*Name of Tank*	*Ownership*	*Area (acre)*	*Village*	*GP*	*Block*
26	Bara Bund	Community	6.7	Anara	Anara	
27	Chimer Bund	Community	3.0		G.P.	Para
28	Bedar Bund	Community	1.5			Para
29	Surjagoria	Govt.	10.0	Palma		Block
30	Dubrajgoria	Community	2.5	Amlatora		
31	Sayer Bund	Community	10.0	Rampur		
32	Sankra Bund	Community	40.3	Sankra	JJ-1 G.P.	
33	Panfalia Bund	Community	8.3	Jhapra		
34	Sing Bund	Community	10.0			
35	Hir Bund	Personal	5.0			
36	Bara Bund	Community	13.7	Jabarra		
37	Lilu Maji Bund	Community	8.0			
38	Chhoto Bund	Community	10.0			
39	Jadu Majhi Bund	Community	2.0	Layara	JJ-2	
40	Bara Bund	Community	4.0		G.P	
41	Pathra Bund	Community	5.0			
42	Gopalgeria Bund	Community	2.0			
43	Mandal Bund	Community	3.5			
44	Siyal Bund	Community	5.0	Asanmoni		
45	Salgeria Bund	Personal	5.0			
46	Chaita Bund	Community	2.0			
47	Arjun Bund	Community	6.5	Ranipur		

(*Contd.*)

Table 3.3 (*Contd.*)

Sl. No.	*Name of Tank*	*Ownership*	*Area (acre)*	*Village*	*GP*	*Block*
48	Lalgeria Bund	Community	1.5	Jorberia		
49	Lombu Bund	Community	10.0			
50	Ekara Bund	Personal	1.5			
51	Bara Bund	Community	7.0	Natundi		
52	Sarak Bund	Community	2.0	Fusrabaid		
53	Lahasyer Bund	Community	21.3	Para	Para G.P.	
54	Parirgeria Bund	Personal	4.0			
55	Ind Sayer Bund	Community	20.0			
56	Sankraha Bund	Community	8.3			
57	Sankra badalgoria	Govt.	0.8			
58	Mahato Bund	Community	6.0	Tentulhiti		
59	Sital Pukar	Community	4.0	Kelahi		
60	Line Choutalageria	Community	4.5	Hariharpur		

Source: Primary (Field) Survey.

tank numbers are 60 in the Purulia district covering 8 GPs and 3 blocks.

In the above, we have given the description of the tanks and the villages in which the tanks are situated. In this chapter, we will give the characteristics of the tanks. We will also analyse the performance and efficiency of the tanks under study in this chapter.

3.2 Tank Classifications

Tanks can be classified on different criteria like ownership pattern, type of irrigation like flow or lift and nature of tanks like old tank or new tank.

Ownership of Tanks: Governance is an important issue in water management. Government, private people and community may be involved in waste management. Panchayat Raj Institutions (PRIs) have also great role to play in water management. The tanks in the state may be grouped under ownership (personal, government and community), type of irrigation (flow, lift and flow-*cum*-lift) and by nature of tanks (old tanks and new tanks). The details of the tanks surveyed in the three districts are given in Table 3.4. Out of the 181 tanks surveyed the ownership-wise distribution of tanks are as follows: Government-32, Personal-69, Community-80. Therefore, the community ownership is highest in our sample followed by personal and government.

Table 3.4: District-wise Numbers of Tanks by Ownership

Districts	*Personal*	*Government*	*Community*	*Total*
Bankura	31	1	20	52
Paschim Medinipur	32	26	11	69
Purulia	6	5	49	60
Total	69	32	80	181

Source: Based on Primary (Field) Survey.

Types of Irrigation: The percentages of tanks by types of irrigation have shown that 62 per cent of the tanks are of dual in nature, i.e. both lift and flow irrigation types while 10 per cent of the tanks are of only lift irrigation and 28 per cent of the tanks are of flow irrigation type (Table 3.5).

Table 3.5: District-wise Percentages of Tanks by Types of Irrigation

Districts	Flow Irrigation	Lift Irrigation	Lift and Flow Irrigation	Total
Bankura	2	17	33	52
Paschim Medinipur	6	14	49	69
Purulia	11	19	30	60
Total	19	50	112	181

Source: Based on Primary (Field) Survey.

Tanks by Nature: We have divided the tanks by their nature into two categories, namely old tanks and new tanks. New tanks have been constructed under the government schemes like NREGS, RSVY, BCW etc. Table 3.6 shows that the distributions of tanks according to their categories, viz. old Tanks = 166 and new Tanks = 15.

Table 3.6: District-wise Numbers of Tanks by Nature

Districts	Old Tank	New Tank	Total
Bankura	49	3	52
Paschim Medinipur	59	10	69
Purulia	58	2	60
Total	166	15	181

Source: Based on Primary (Field) Survey.

Among the three types of classification, the ownership of the tanks (Personal, Government and Community) is more relevant as most of the interventions are addressed using this classification. Hence, this classification is used for further discussion and analysis of the survey data under this study.

We have divided the discussion on the characteristics of tank irrigation in three dimensions – extent of irrigation, conditions of tanks and efficiency of tanks.

3.3 Extent of Irrigation

Net Cultivated Area: Average actual gross cultivated area of the tanks in the sample is 56.96 acres of which acreage for different crops are: Aman paddy-50.7 acre, Boro paddy-0.93 acre, others-5.31 acres. Average net cultivated area under different crops for the classifications is presented in Table 3.7. As the table reveals Aman paddy is the main crop in the study area.

Table 3.7: Category-wise Average Actual Net Cultivated Area for the Tanks (in Acres)

Category	*Aman*	*Boro*	*Other*	*Total*
Personal	41.15	0.89	4.01	46.05
Government	61.42	2.52	5.85	69.79
Community	54.69	0.32	6.21	61.22
Total	50.72	0.93	5.31	56.96

Source: Based on Primary (Field) Survey.

Caste of Beneficiaries: It is important to examine the land ownership patterns in the tank command areas in terms of caste and land holding size which are regarded as crucial determinants of collective action. The percentages for different castes for the whole sample are as follows: General-30 per cent, OBC-30 per cent, SC-22 per cent, ST-18 per cent. Therefore, the backward caste for the whole sample consists of 60 per cent. Caste composition under three classifications is presented in Table 3.8. It may be seen that percentage of ST is highest for new tank under classification 3 which reflects the government policy.

Table 3.8: Category-wise Percentages of Caste of Beneficiaries under Tank Command Area

Category	*Gen*	*OBC*	*SC*	*ST*	*Total*
Personal	32	39	14	15	100
Government	31	29	18	22	100
Community	29	25	28	18	100
Total	30	30	22	18	100

Source: Based on Primary Survey.

Beneficiaries per Tank by Land Distribution: The average number of beneficiaries from tank irrigation for different land distributions has been given in Table 3.9. The average number of beneficiaries is calculated as 67 per tank for the whole sample. The land class-wise average number of beneficiaries for the whole sample are as follows: up to 1 acre–32, (1-2) acre–22, (2-4) acre–8, (4-5) acre–2.7, (5-6) acre–1.2, (6-7) acre–0.9 and above 7 acre–0.5. Obviously beneficiaries are mostly within the land class of (0-1) he as is the land distribution. Under classification 1 average number of beneficiaries is highest for community owned tanks.

Table 3.9: Category-wise Average Numbers of Beneficiaries per Tank by Land Distribution (Acres) under Tank Command Area

Category	*Up to 1*	*1-2*	*2-4*	*4-5*	*5-6*	*6-7*	*Above 7*	*Total*
Personal	27.6	20.1	5.1	1.6	0.2	0.3	0.1	55.0
Government	29.2	25.1	11.6	2.2	1.2	0.8	0.8	70.7
Community	37.2	22.4	9.1	3.9	1.9	1.4	0.6	76.5
Total	32.1	22.0	8.0	2.7	1.1	0.9	0.5	67.2

Source: Based on Primary (Field) Survey.

Cropping Pattern: We have collected data for the sample tanks on the crops that are grown in the tank command area. Twelve crops are seen to be grown in the tank command area of our sample tanks. The crop-wise percentage of tanks are given in Table 3.10: Aman–100 per cent, Boro–7 per cent, Wheat–40 per cent, Potato–20 per cent, Seasame–17 per cent, Mustard–44 per cent, Sunflower–1 per cent, Vegetables–50 per cent, Onion–3, Cauliflower–2 per cent. Gourd-1 per cent and Sugarcane–1 per cent. Therefore, Aman is produced in each tank command area, followed by vegetable, wheat, mustard and potato.

Table 3.10: Category-wise Percentages of Tanks by Crops in Tank Command Area

Category	*Aman*	*Boro*	*Wheat*	*Mustard*	*Potato*	*Sun-flower*	*Sugar-cane*	*Seasame*	*Vegetable (Gourd, Cauliflower, Onion, Tomato etc.)*
Personal	100.0	5.8	21.7	43.5	15.9	0.0	0.0	20.3	27.5
Government	100.0	12.5	37.5	56.3	25.0	0.0	0.0	28.1	28.1
Community	100.0	5.0	55.0	40.0	22.5	1.3	2.5	8.8	56.3
Total	100.0	6.6	39.2	44.2	20.4	0.6	1.1	16.6	40.3

Source: Based on Primary (Field) Survey.

Price of Lands: We have collected data about the price of irrigated and non-irrigated land. The difference between the two reflects the importance of irrigation. The average price of irrigated land for the whole sample is Rs. 55.8 thousands per acre, the average price of non-irrigated land is Rs. 32.5 thousands per acre and the percentage difference is 81.6. The highest price difference is for community owned tank (Table 3.11).

Table 3.11: Category-wise Average Price of Lands in the Tank Command Area

Category	*Price of Irrigated land*	*Price of Non-Irrigated land*	*Percentage Difference in Price*
Personal	44.1	28.2	77.2
Government	56.3	36.6	59.1
Community	65.7	34.5	94.3
Total	55.8	32.5	81.6

Source: Based on Primary (Field) Survey.

Temperature and Water Level: Temperature and groundwater level as collected from the respondents have presented in Table 3.12. The temperature is collected for the summer season. The groundwater level is collected for all the three seasons. These are the averages of field level data. In the summer, groundwater obviously declines. The average groundwater level for the whole sample is 55 ft for summer, 10 ft for rainy season, 38 ft for winter season. The average groundwater level for the sample of Bankura is 47 ft for summer, 9 ft for rainy season, 36 ft for winter season. The average groundwater level for the sample of Paschim Medinipur is 47 ft for summer, 9 ft for rainy season, 36 ft for winter season. The average groundwater level for the sample of Purulia is 55 ft for summer, 11 ft for rainy season, 36 ft for winter season.

Table 3.12: Blocks and District-wise Average of Present Temperature (°C) in Summer and Ground water Level (Ft)

Districts	*Temperature (°C) In summer*	*Summer (GWL) (ft)*	*Rainy (GWL) (ft)*	*Winter (GWL) (ft)*
Bankura	42	47	9	36
West Medinipur	37	62	9	41
Purulia	41	55	11	36
Total	40	55	10	38

Source: Based on Primary (Field) Survey.

3.4 Tank Conditions

Category-wise Condition of the Tanks: We have assessed tank condition on 5-point scale with scaling values as follows: Very good - 5, good - 4, medium - 3, bad - 2 and very bad - 1. We have taken five parameters as indicators of quality of tank, namely the condition of bund, system of entering water, exit system of water, channel maintenance and desiltation. The average scale values for the above mentioned 5 parameters for each classification are given in Table 3.13. The scale values suggest that channel maintenance and desiltation are not satisfactory.

Table 3.13: Category-wise Condition of the Tanks Based on the Scale Values

Category	Bund Condition	Enter Water	Exit Water	Channel Maintenance	Desiltation
Personal	3.26	3.71	3.74	1.41	1.19
Government	3.22	3.50	3.56	1.41	1.16
Community	3.16	3.56	3.31	1.53	1.14
Total	3.21	3.61	3.52	1.46	1.16

Source: Based on Primary (Field) Survey.

Average Area and Depth of the Tanks: In Table 3.14, we have given the average of the area and depth of the sample tanks in category-wise. Here we see that the average area and depth are 7.8 acre and 8.7 ft in Personal Category tank respectively, the average area and depth are 6.1 acre and 12.6 ft in Government Category tank respectively and the average area and depth are 7.3 acre and 10.5 ft in Community Category tank respectively. The grand average of the three categories for tank area is 7.2 acre and the depth is 10.2 acre respectively.

Table 3.14: Category-wise Average Area (Acre) and Depth (ft) of the Tanks

Category	Area (in Acre)	Depth (ft)
Personal	7.8	8.7
Government	6.1	12.6
Community	7.3	10.5
Total	7.2	10.2

Source: Based on Primary (Field) Survey.

Water Availability: Water in most of the tanks get dried out in the summer season. We have first calculated average availability of water measured in terms of months. As the water is used for some other purposes like fishery and bathing, the availability of water for irrigation is less than average availability. We have calculated water availability and water uses for irrigation for the sample tanks by the number of months (Table 3.15). Here we have seen that the average water availability for the whole sample is 10.3 months and the water used for irrigation purposes for 6.2 months on average.

Table 3.15: Category-wise Average of Water Availability (months), Water Uses (months) for Cultivation and Bathing

Category	*Total Water Availability (months)*	*Water availability for Cultivation (months)*
Personal	9.8	5.2
Government	10.1	6.2
Community	10.9	7.0
Total	10.3	6.2

Source: Based on Primary (Field) Survey.

Institutions for Tank Water Management: It is now increasingly being felt that sustainable irrigation management is not possible without decentralized participatory irrigation management. Andhra Pradesh is the first state to form WUA in 1997. The main objectives for formation of WUA are—to realize the maximum irrigation potential, improving the efficiency of existing irrigation system, ensuring equitable and reliable water supply and better water resource management by the users themselves. Though this is often discussed at the policy level, it is not done at the implementation level. Water Users' Association can be formal or informal. In the first case institutions are formalized through legislation. In our study area we have not observed any formal institutions. All the users associations are informal. Generally informal institutions have roots in local communities and are embedded in with the existing custom, traditions, norms and beliefs (*Reddy et al.* 2008). As our analysis shows the presence of association has impact on land utilization of farmers.

We have collected information about whether there is association or committees for the management of the tanks. Table 3.16 shows that out of the 181 tanks only 44 tanks (24%) are reported to have informal type associations and 137 tanks have no associations. The name of committees is different in different places: *Solo Ana,* Club, Gram Panchayat, *Gram Unnayan Committee*, *Sabar Committee*, and *Ansidar*. Out of the 42 committees, 23 committees are named as *solo ana* committees (Table 3.16). This indicates that water management through participation has not been implemented in the study area. There are several reasons for this. For personal tanks, the owner of the tank does not like his tank to be managed by other persons or association. From field study it is revealed that some tank owners are averse to apply for NREGS funds on the fear that some other persons may interfere or may get access to the water of their private tanks. Some associations have been evolved out of family fragmentation. The persons having their land nearby get advantage of irrigation water. Government-managed tanks often use the water for fish production. The community-managed tanks often use the sales proceeds from fish and water for community purposes like creating community assets. Less collective effort is seen in maintaining the tanks. Though water users association has been running well in many cases in well irrigation in West Bengal, it is not so for tank irrigation. The main hindrance in forming this is the lack of awareness and motivation of people in building collective effort to manage the tanks. People are more interested in getting government fund, instead of generating internal resources for maintaining the funds.

Table 3.16: Category-wise Numbers of Tanks by Name of the Associations

Category	*Solo Ana*	*Club*	*Gram Panchayat*	*Gram Unnayan Committee*	*Sabar Committee*	*Ansidar*	*Total*
Personal	4	2	0	0	0	3	9
Government	0	0	1	1	1	0	3
Community	19	1	3	1	0	6	30
Total	23	3	4	2	1	9	42

Source: Based on Primary (Field) Survey.

Average Expenditure Construction/Re-excavation of Tanks: In Table 3.17, we have presented that the average expenditure of construction/re-excavation of tanks/bunds upon the available data from our field survey. Here we see that the average expenditure per tank is Rs. 2.4 lakh. Under this classification the highest average expenditure is for community-owned tank. We also see that the average expenditure per gross cropped area (GCA) is highest in community-owned tank.

Table 3.17: Category-wise Average Expenditure (Rs.) of Construction/Re-excavation per Tank

Category	*Average Expenditure of Construction/Re-excavation of Tanks (Rs. in Lakhs)*	*Average Expenditure of Construction/Re-excavation of Tanks (Rs. in '000)/GCA*
Personal	1.4	2.8
Government	2.1	2.8
Community	3.6	7.5
Total	2.4	5.5

Source: Based on Primary (Field) Survey.

We have collected some information about scheme-wise expenditure under different programmes. We mention here average expenditure for these schemes:

1. *RSVY of Jhargram, Paschim Medinipur (From Jhargram Agri-Irrigation)*: The average expenditure per scheme is calculated as Rs. 3,15,059.
2. *RSVY of Bankura (From Bankura Agri-Irrigation)*: The average expenditure per scheme is calculated as Rs. 5,53,311.
3. *RSVY of Purulia (From Purulia Agri-Irrigation)*: The average expenditure per scheme is calculated as Rs. 4,72,154.
4. *BCW of Bankura (From Bankura Agri-Irrigation)*: The average expenditure per scheme is calculated as Rs. 4,72,154.
5. *BCW of Purulia (From Purulia Agri-Irrigation)*: The average expenditure of construction is calculated as Rs. 4,54,891 and the average cost of re-execution is Rs. 6,46,301.
6. *Bankura Jilla Parisad (Immovable Property)*: We have collected the list of bundhs and tanks under Zilla Parishad of Bankura. The average area of 104 schemes is 9 acres,

Problems in Tank Irrigation: We have collected and compiled different problems about tank irrigation or tank irrigation related problems. In the tank questionnaire there was a question—"What are the problems of tank irrigation?" No options were given with the question. This Table 3.18 is prepared from the answers they had given.

Major problems that came out are less irrigation channels, less depth of the tank, less re-excavation, lack of maintenance.

Table 3.18: Problems of the Tanks

Problems	*Nos. of Tanks*	*Percentages*
Less Depth	85	47
No Irrigated Channel	77	43
Irregularities of re-excavation	60	33
No re-excavation	50	28
Bad bund condition	36	20
Water use for bathing	29	16
Less area of the tank	25	14
Irregularities of maintenance of field channel	20	11
Improper use of allotment	19	10
Excessive water use in Aman cultivation	18	10
No desiltation system	10	6
Less/Irregularities of scheme allocation by Government	9	5
Bad Outlet	8	4
No Outlet/'*Magra*'	7	4
Bad quality of soil	8	4
Bad water distribution system	6	3
No connection between River and Tank	6	3
No link between tank and other sources of irrigation	5	3
No afforestration	5	3
Bad Inlet	4	2
Irregularity of maintenance	3	2
Unsatisfactory role of Panchayat	3	2
Leakages of Bund	4	2
No lock gate in tank	4	2
Absence of WUA	4	2
Less water storing capacity	1	1
Misuse of water	1	1
Beneficiary not interested for cultivation	2	1
Bad condition of catchment Area	1	1
No electricity	1	1
Less height of bund	2	1
Encroachment in tank command Area	1	1
No protection of the overload of water	1	1
Problems of water carrying system	1	1
No security guard	2	1
Improper plan	1	1

Source: Based on Primary (Field) Survey.

3.5 Performance of Tanks

We have done the efficiency analysis of tanks in three ways—(i) tank productivity analysis, i.e.. productivity of the tank command area, (ii) efficiency measurement through DEA, and (iii) land utilization analysis.

3.5.1 Productivity of Tanks under Study

In Table 3.19 we have presented the gross cropped area per tank, total value of agricultural production, and fishery value. The total value of production is higher in case of government category, even though fishery value is lower in this category.

Table 3.19: Category-wise Average Gross Cultivated Area (in acre) Per Tank

Category	Cultivated Area (in acre)	Total Value from Cultivation (Rs. in lakh/acre)	Fishery (Rs. in lakh/acre)
Personal	50.0	5.9	0.2
Government	71.5	8.5	0.1
Community	62.7	6.6	0.2
Total	59.4	6.6	0.2

Source: Based on Primary (Field) Survey.

In Table 3.20 below we have presented results of productivity of Aman and Boro paddy in quintals per acre for different categories of tanks. We have already noted in addition to aman and boro paddy, farmers produce other crops like vegetables. The average productivity in the tank command area is calculated as Rs. 30.4 thousand per acre. Average productivity is seen to be highest for the community category with Rs. 33.1 thousand per acre. The ANOVA test results show that the average productivity of the tank command area for the three categories is different at the significance level of 17 per cent.

Table 3.20: Category-wise Average Productivity in Tank Command Area

Category	Aman (Q/Acre)	Boro (Q/Acre)	Average Productivity (Rs. '000./Acre)	F value	p value
Personal	15.6	17.2	26.7	1.79	0.17
Government	16.5	15.2	31.8		
Community	14.0	16.3	33.1		
Total	15.0	16.3	30.4		

Source: Based on Primary (Field) Survey.

Regression Analysis: In order to have the exact impact of the influencing variables on the tank productivity, a regression analysis was done using the following multiple linear regression model.

$$\text{TOTPODY} = a + b_0\,\text{TANKCOND} + b_1\,\text{DVEG} + b_2\,\text{DASSOCIA} + b_3\,\text{DFISHERY} + b_4\,\text{VOLUME} + b_5\,\text{DGOVT} + b_6\,\text{DCOMMUN}$$

Where,

TOTPODY = productivity at tank level (Rs./Acre)

TANKCOND = score value of tank condition

DVEG = vegetable dummy (= 1 if vegetable is cultivated in the tank command, 0 = otherwise)

DASSOCIA = whether association is present for the tank (=1 if present, 0 = otherwise)

DFISHERY = whether fishery is present in the tank (=1 if present, 0 = otherwise)

VOLUME = volume of the tank (in acre-feet)

DGOVT = whether the tank is owned by the Government (= 1 if yes, 0 = otherwise)

DCOMMUN = whether the tank is owned by the Community (= 1 if yes, 0 = otherwise)

The results are reported in Table 3.21.

Table 3.21: Estimated Regression Results

	Unstandardized Coefficients	*t*	*Sig.*
	B		
(Constant)	18848.208	1.116	.266
TANKCOND	3278.184	.516	.606
DVEG	20656.816	4.104	.000
DASSOCIA	182.264	.033	.974
DFISHERY	–8750.413	–1.711	.089
VOLUME	10.155	1.417	.158
DGOVT	13844.134	2.165	.032
DCOMMUN	3310.168	.640	.523

Source: Based on Primary (Field) Survey.

The results show that fishery dummy has negative co-efficient. This is because if there is fishery then availability of water for irrigation decreases. The positive significant vegetable dummy indicates that tanks with vegetables have higher productivity in the tank command area thus confirming that crop diversification is a good alternative for maximizing the income from the tank irrigation. Also for every feet increase in the depth of tank water, the production level will increase by Rs. 10/acre and for a command of say 100 acres, this will be about Rs. 5000. The ownership by Government clearly indicates the possibilities of higher income as indicated by the higher coefficients of the variables. Government-owned tanks have significant impacts on productivity. The significance of the volume of the tank indicates the importance of the depth of the tank. The marginal impact of Government-owned tanks is higher than community owned tanks. As most of the associations are informal and not well organized, it has less impact on productivity.

3.5.2 Efficiency Analysis of Tanks using DEA

Water resources management reform occupies a prominent place in agricultural policy due to the scarcity of water resources and the associated issue of management efficiency. The objective of this section is to assess the efficiency of the irrigation tanks in our sample. We have used data envelopment analysis (DEA) for judging the efficiency of the tanks. The methodology was developed by Charnes, Cooper and Rhodes based on M.J. Farrel's contribution to productive efficiency. It is linear programming based technique for measuring the performance efficiency of a decision making unit (DMU).

DEA models vary in the following ways: according to scale conditions these are constant or variable; according to the models orientation these models may be input or output oriented model and to the mathematical form of the objective function may differ like piecewise linear, log-linear or Cobb-Douglas forms.

DEA models differ in the emphasis of their orientation (input oriented or output oriented) and on the assumption of returns to scale (CRS or VRS). Input-oriented DEA models tries to maximize the proportional decrease in input variables while remaining within the envelopment space (production possibility set), but output-oriented DEA models maximize the proportional increase in the output variables while remaining within the envelopment space. In other words input oriented technical efficiency is based on the question that how much output quantities be proportionally reduced without changing the output quantities produced. On the other hand, ouput oriented DEA is based on the question that how much output quantity be expanded without altering the input quantities used. Thus the possible improvement of an inefficient DMU could be achieved by either focusing on a proportional input reduction or proportional output augmentation or both simultaneously. Under VRS the envelopment surface presents convexity as a consequence of the constraint ($1\lambda = 1$) in the model. It ensures that the inefficient DMUs are measured in comparison with those efficient DMUs on the efficient frontier operating at the same scale size as themselves, the efficiency measured is 'pure technical efficiency' without considering the impact of the operating scale size of each DMU. The input-oriented VRS DEA model adopted in this study for input saving efficiency is an input-oriented VRS model as follows (*Gang and Flemingham 2002*).

$$\text{Min } (\theta, \lambda, S^{+}, S^{-})$$
$$TEc = \theta - \varepsilon\,(1s^{+} + 1s^{-})$$
$$\text{s.t. } Y\lambda - s^{+} = Y_0$$
$$\theta X_0 - X\lambda - s^{-} = 0$$
$$1\lambda = 1$$
$$\lambda, s^{+}, s^{-} \geq 0$$

Where TEc is the cost of technical efficiency score and the variable q is the (proportional) reduction applied to all inputs of DMU_0 (the DMU being evaluated) to improve efficiency, λ is the value of constructing a composite unit with inputs $Y\lambda$ and outputs $X\lambda$ outperforming DMU_0, ε allows the minimization over θ to pre-empt the optimization involving the slack variables s+ and s^{-}, the variables s^{+} and s^{-} represent the excess inputs and output slacks and are the distance between an inefficient DMU (X_0, Y_0) and its projected point on the envelopment surface of efficiency frontier. As can be seen, an input-oriented BCC model focuses on the maximal movement towards the frontier through a proportional reduction of inputs, given the output level.

Following the DEA methodology for input oriented VRS model we have derived the efficiency score of each tank (*Coelli 1996*). Here the inputs have been taken as the area and depth of the tanks and output of the tanks has been taken as gross actual command area of the tanks. We have first derived the efficiency scores of each tank and then we have calculated the average efficiency score of tanks for the different categories of tanks. In the following Table 3.22, we have shown the efficiency score of the tanks. The efficiency of tanks for different categories of tanks assuming constant returns to scale are as follows: Personal–0.65, Government–0.59 and community–0.56.

Table 3.22: Category-wise Performance of the Tanks by Scale

	Average Efficiency	*Standard Deviation*
Personal	0.30	0.21
Government	0.43	0.27
Community	0.34	0.22

We have observed that efficiency scores of tanks differ widely across tanks. The difference in efficiency scores may arise due to physical characteristics of the region, socio-economic factors or water management differences. We have tried to find the determining factors for efficiency scores of tanks. The F value shows the significance of the overall regression.

ANOVA

Model		*Sum of Squares*	*Df*	*Mean Square*	*F*	*Sig.*
1	Regression	1.927	5	.385	11.232	.000
	Residual	6.005	175	3.432E-02		
	Total	7.932	180			

a Predictors: (Constant), COMMUNIC, FISHERY, IPBNIP, FARFARME, PAKKA

b Dependent Variable: EFFSCORE

In the Table below, we have reported the determining factors of efficiency scores. The factors are COMMUNIC (Communication score in the village with scale 1 to 5), FISHERY (whether there is fishery in the tank), IPBNIP (Irrigated price by non-irrigated price in the tank command area), FARFARME (Far farmer gets less water), PAKKA (Whether there is pakka road in the village). The results show that FARFARME is positively related to efficiency, i.e. inequality in the distribution of water is leading to increase in efficiency. The FISHERY is negatively related to efficiency, i.e. the tanks that have fishing have lower efficiency because these tanks irrigate less. The efficiency score is negatively related IPINB, i.e. relative high price of irrigated land have lower efficiency. The efficiency score is positively related with PAKKA and COMMMUNICATION.

Table 3.23: Regression Result of Efficiency of Tanks

	Model	*Unstandardized Coefficients*	*t*	*Sig.*
		B		
1	(Constant)	.428	5.108	.000
	FARFARME	7.125E-02	1.928	.055
	FISHERY	–9.247E-02	–2.482	.014
	IPBNIP	–5.444E-02	–2.127	.035
	PAKKA	3.371E-02	5.740	.000
	COMMUNIC	8.756E-02	3.551	.000

a Dependent Variable: EFFSCORE (Efficiency Score of Tanks)

3.5.3 Land Utilisation in the Tank Command Area: Proportion of Boro and Other Crops (Non-Aman) Cultivation in Tank Command Area

The performance of the tanks can also be judged whether the second crop is cultivated in the tank command area. The main crop in all the tank command areas is Aman crop, which is rainfed crop. We have collected data about the amount of cropped area of Aman and Boro and other crops. Then we have calculated the proportion of boro and other crops area (i.e. non-aman cropped area) to the total cropped area. We used the Tobit regression model (*Hill Griffiths Lim 2008*) to search for the significant factors affecting the probability of growing other crops besides aman. The results (given in the table below) show that quality of the tank as judged by the tanks' condition (COND) (described earlier), presence of association, presence of fishery and depth of the tank have positive impact on the proportion of other crop cultivation, i.e. intensive utilization of land in the tank command area. Here we see that presence of fishery in the tanks leads to higher utilization of tank command area. The ownership pattern of the tanks does not have much impact on the utilization.

There should be more emphasis on crop diversification. Associations should be more formalized so that these can play greater role in enhancing the productivity of the tank command area. There should be more regulations in the maintenance of tanks. Desiltation can enhance the depth of the tank so that the volume of water retained in the tank can be enhanced. Tank conditions should be improved so that productivity of the tanks could be enhanced. Since Government tanks indicate higher efficiency, the possibilities of improving the tanks under other ownership should be examined. The productivity also depends on other factors like soil conditions, intensity of use of inputs like fertilizer and pesticide which are not taken into consideration in this study.

Table 3.24: Tobit Regression Results

	Coefficient	*Std. Error*	*z-Statistic*	*Prob.*
C	–0.096692	0.072206	–1.339114	0.1805
ASSOCIATION	0.029387	0.027308	1.076137	0.2819
CONDITION	0.009831	0.005481	1.793628	0.0729
DEPTH	0.002468	0.002285	1.080188	0.2801
FISHERY	0.036653	0.029787	1.230486	0.2185
DGOV	0.011002	0.032998	0.333419	0.7388
DCOMMUNITY	0.014498	0.029212	0.496324	0.6197

Summary of the Chapter

In this chapter, we have presented the tank study based on primary survey. We have presented the tank characteristics with ownership classification—Personal, Government and community. The characteristics include—extent of irrigation, conditions of tank and irrigation efficiency. Extent of irrigation includes category-wise net cultivated area per tank, number of beneficiaries etc. Tank conditions include different dimensions of tank conditions. We have also discussed institutional arrangement for the involvement of users like water users association. The efficiency of the tanks have been analysed from the three aspects: productivity of tank command area, efficiency on the basis of DEA and the cropping intensity. The determinants of these three types of efficiencies have been found out.

Chapter 4

Household Study: Primary Survey

In the previous chapter, i.e. Chapter 3 we have described about different characteristics of sample tanks. In this chapter we will present the study on households in tank command areas. The study includes households' demographic and socio-economic characteristics, perception of the households about tank quality, irrigation cost, production cost and profit of different crops in the tank command area.

4.1 Households Characteristics

4.1.1 General Information

Members of Households: This section will provide the different characteristics of surveyed households. We have already described that 533 households have been surveyed in the three districts of West Bengal. We have selected the regions of the districts, which are dry in nature. Table 4.1 gives the percentage of male and female members of the households. Here we see that the percentage of female is 46 and the percentage of male is 54 respectively.

Table 4.1: Tank Category-wise Percentage of Male and Female Members of Households Surveyed

Category of the Tank	*Male*	*Female*	*Total*
Government	54	46	100
Personal	53	47	100
Community	55	45	100
Total	54	46	100

Source: Based on Primary (Field) Survey.

Family Size of the Households: We have presented to tank category-wise family size in Table 4.2. The category-wise family sizes as calculated on the basis of the primary survey are as follows: Government–5.6, Personal–6.5 and Community–6.7. The family size in the category of community is higher than the grand average.

Table 4.2: Tank Category-wise Family Size of the Households

Category of the Tank	*Male*	*Female*	*Total*
Government	3.0	2.6	5.6
Personal	3.4	3.1	6.5
Community	3.7	3.0	6.7
Total	3.5	3.0	6.4

Source: Based on Primary (Field) Survey.

Caste Categories of the Households: We have presented the caste-wise distribution of households under our study for different tank categories in Table 4.3. It shows that the caste-wise distribution patterns of households for the whole are as follows: General–37 per cent, OBC–33 per cent, SC–11 per cent, ST–19 per cent. Therefore, our sample households are dominated by backward castes. As for tank category percentage of ST is higher for community.

Table 4.3: Tank Category-wise Percentage of Households by Castes

Category of the Tank	*General*	*OBC*	*SC*	*ST*	*Total*
Government	35	29	7	28	100
Personal	37	29	14	21	100
Community	38	37	11	14	100
Total	37	33	11	19	100

Source: Based on Primary (Field) Survey.

Types of Ration Card: We have collected data whether the households belong to APL, BPL and Antyodaya. In Table 4.4, we see that the percentages are respectively as follows: APL–66 per cent, BPL–30 per cent, Antyodaya–4 per cent. We have observed that these categories do not always reflect the true economic condition of the families. Some families have argued that they have been deprived of BPL cards.

Table 4.4: Tank Category-wise Percentages of APL, BPL and Antyodaya Families

Category of the Tank	*APL*	*BPL*	*Antyadaya*	*Total*
Government	72	22	6	100
Personal	65	31	4	100
Community	65	32	3	100
Total	66	30	4	100

Source: Based on Primary (Field) Survey.

Age and Education of Head of the Families: The head of the household plays key role in taking various household decisions. We have collected data about the age and education level of the head of the household. The average age (year) of the head of the households for three tank categories are as follows (in Table 4.5): Government–47 years, Personal–49 years and Community–50 years whereas the average age of the head of the households of the whole sample is calculated 50 years. The average education level attained by the head of the household are as follows: Government–class 5, Personal–class 5 and Community–class 6. The average education of the head of the whole sample is calculated as class 5.

Table 4.5: Tank Category-wise Average of Age (Year) and Education (Class) of Head of the Families

Category of the Tank	*Average Age of the Head of the Family*	*Average Education of the Head of the Family*
Government	47	5
Personal	49	5
Community	50	6
Total	50	5

Source: Based on Primary (Field) Survey.

Age of the Members of the HHs: The age distributions of the family members of the surveyed households are presented in Table 4.6. The age group-wise distribution of family members for the whole is as follows: 0-6 yrs–12 per cent, 7-14 yrs–15 per cent, 15-18 yrs–8 per cent, 19-25 –14 per cent, 26-60 yrs–46 per cent, 61-65 yrs–3 per cent, above 65 yrs.–2 per cent.

Table 4.6: Tank Category-wise Percentage Distribution of Members of Families Surveyed by Age Groups

Category of the Tank	0-6	7-14	15-18	19-25	26-60	61-65	Above 65	Total
Government	11	15	7	20	44	2	1	100
Personal	11	16	8	12	48	3	2	100
Community	12	15	8	13	46	3	2	100
Total	12	15	8	14	46	3	2	100

Source: Based on Primary (Field) Survey.

Education Level of the family members: The detailed tank categories-wise percentage distribution of the members of the sample households by their education levels are presented in Table 4.7. The percentage of family members for different education levels are as follows: child–7.9 per cent, illiterate–13.3 per cent, primary–26.7 per cent, madhyamik pass–42 per cent, higher secondary–5.9 per cent, graduation–3.8 per cent, above graduation–0.3 per cent.

Table 4.7: Tank Category-wise Percentages Distribution of Family Members by Education Level

Category of the Tank	Child	Illiterate	Primary	MP	HS	Graduation	Above Graduation	Total
Government	8.6	15.4	32.6	34.0	5.0	4.4	0.0	100
Personal	7.9	12.5	30.2	41.6	5.2	2.4	0.3	100
Community	7.8	13.3	22.8	44.5	6.6	4.7	0.3	100
Total	7.9	13.3	26.7	42.0	5.9	3.8	0.3	100

Source: Based on Primary (Field) Survey.

Occupations of the members of the households: We have also calculated the occupational pattern of households for different tank categories. The occupational pattern of the households for the whole sample are as follows (in Table 4.8): cultivation–30 per cent, business–4 per cent, service–2 per cent, student–26 per cent, housewife–13 per cent, housewife with cultivation–15 per cent, old person–7 per cent, children–9 per cent, homework–1 per cent. It is obvious that the main source of income is cultivation. It is also to be noted that a major portion of housewives is involved in cultivation.

Table 4.8: Tank Category-wise Percentage of HH Members by Different Occupations

Category of the Tank	Cultivation	Business	Service	Student	House Wife	House Wife+ Cultivation	Home Work	Old+ Child	Total
Government	30	3	3	23	12	17	2	11	100
Personal	32	2	1	25	12	17	1	9	100
Community	29	4	2	27	14	13	1	10	100
Total	30	4	2	26	13	15	1	9	100

Source: Based on Primary (Field) Survey.

Types of houses: The tank category-wise distributions for different types of houses of the sample households are presented in Table 4.9. Out of 533 households 65 per cent households have Kachha (made of mud) houses. As has been given in the Table Kachha (made of mud) houses are dominant for all the categories.

Table 4.9: Tank Category-wise Percentage of Households having Different Types of Houses

Category of the Tank	Kachha	Mixed	Pakka	Total
Government	68	16	15	100
Personal	66	12	22	100
Community	66	12	22	100
Total	65	13	22	100

Source: Based on Primary (Field) Survey.

Types of assets: We have collected data for assets by each household, i.e. whether they possess the assets like TV, Radio, Cycle, Motor Cycle, Bullock cart, and Telephone. The percentages of households for each asset are presented in Table 4.10. The percentages of households for each asset of the sampled households are as follows: TV–36 per cent, Radio–9 per cent, Cycle–94 per cent, Motor Cycle–13 per cent, Bullock Cart–41 per cent, Land Phone–1 per cent, and Mobile–51 per cent.

Employment and Income from NREGS: We have also calculated tank category-wise average of working days, average income and percentage of job card holders of families from NREGS which is given in Table 4.11. The percentages of job cardholders are as follows: Government–80 per cent, Personal–81 per cent and

Table 4.10: Tank Category-wise Percentage of HHs having Various Type of Assets

Category of the Tank	*TV*	*Radio*	*Cycle*	*M Cycle*	*Bullock Cart*	*Telephone (land)*	*Mobile*
Government	31	9	87	15	27	0	39
Personal	30	10	93	7	38	1	48
Community	42	9	96	16	48	0	57
Total	36	9	94	13	41	1	51

Source: Based on Primary (Field) Survey.

Community–71 per cent. The average employments per family are as follows: Government–29 days, Personal–26 days and Community–26 days. The average employment per family per annum is 26 days for the whole sample. The Average annual income per family from NREGS is Rs. 1836 for the whole sample.

Table 4.11: Tank Category-wise Average of Working Days, Average Income from NREGS and Percentage of Job Card Holders of Families (2008-09)

Category of the Tank	*% of NREGA Job Card holders*	*Average no. of days of employment in NREGA*	*Average income (Rs./year) from NREGA*
Government	80	29	1969
Personal	81	26	1831
Community	71	26	1790
Total	76	26	1836

Source: Based on Primary (Field) Survey.

Per Capita Working Days: In Table 4.12 we have presented tank category-wise per capita working days for working ages (19-60) in different sectors. Here we see that in all three categories most of the people are involved in cultivation. The percentages of these are 72 in Government category, 81 in Personal category and 89 in Community category. Out of the average employment days of 135 per year per worker, the number of days of employment for different occupations are as follows: cultivation-84, business-13, service-10, NREGA-4, non-agriculture-25.

Table 4.12: Tank Category-wise Per Capita Working Days in Different Sectors in a Year (Age Group 19-60)

Category of the Tank	*Cultivation*	*Business*	*Service*	*NREGA (Family)*	*Non-Agriculture*
Government	72	14	13	5	18
Personal	81	9	5	4	25
Community	89	15	13	3	26
Total	84	13	10	4	25

Source: Based on Primary (Field) Survey.

Burden of Dependency: We have also calculated the dependency burden of the households. Here we see that the percentage of total dependents (0-14 and above 60) to total members is 32 per cent and the percentage of working age members (15-60) to total members is 68 per cent. We have calculated the dependency burden of the household as population within 0-14 yrs. and population within 60+ yrs/persons within the age group 15-60 yrs which are shown in Table 4.13.

Table 4.13: Tank Category-wise Burden of Dependency of Dependent Persons on Working Ages (15-60 Years)

Category of the Tank	*% of Total Dependants (0-14 and above 60) to total members*	*% of Working age members (15-60) to total members*	*Dependency Ratio*
Government	30	70	0.43
Personal	32	68	0.47
Community	32	68	0.48
Total	32	68	0.47

Source: Based on Primary (Field) Survey.

Monthly Income and Expenditure and engaged in cultivation: We have collected data about average monthly income and expenditure of family members and also the average number of months engaged in cultivation from our field survey. These are shown in Table 4.14. Here we see that the average monthly income and expenditure of the family members are Rs. 4.9 thousand and Rs. 2.9 thousand respectively in whole sampled families. In most of the villages, there is little opportunity of alternative employment. In Table 4.14, we have seen that most of the people are engaged in cultivation. The average engagement in agriculture is 8 months for all the districts.

Table 4.14: Tank Category-wise Average Monthly Income and Expenditure and Engaged in Cultivation

Category of the Tank	*Average Monthly Income (Rs. in '000) of the HHs*	*Average Monthly Expenditure (Rs. in '000) of the HHs*	*Average no. of months engaged in cultivation*
Government	5.0	2.8	8
Personal	3.4	2.3	8
Community	6.0	3.3	8
Total	4.9	2.9	8

Source: Based on Primary (Field) Survey.

Size Class of Land Holding: In Table 4.15, we have presented the percentage of households surveyed by size of land holding. Out of 533 households surveyed 55 per cent households belonged to marginal, 31 per cent households belonged to small, 10 per cent households belonged to medium and 5 per cent households belong to big. From this table we can tell that the percentage of marginal land holding is highest in Government categories.

Table 4.15: Tank Category-wise Percentage of Households Surveyed by Size Class of Land Holding

Category of the Tank	*Marginal*	*Small*	*Medium*	*Big*	*Total*
Government	61	28	11	0	100
Personal	56	33	9	2	100
Community	52	31	10	8	100
Total	55	31	10	5	100

Source: Based on Primary (Field) Survey.

Pattern of land holding to Total agricultural land: We have classified land into five categories—own, barga, patta, lease-in and lease-out land, which are shown in Table 4.16. Our calculation reveals that 97.9 per cent of land holding is own land, 1.3 per cent of land holding is *Barga* land, 0.1 per cent of land holding is *Patta* land and 1.1 per cent of land holding is Lease in land. On the other hand the percentage of Lease out land is 0.4. There is not so much tank category variation in the land holding pattern.

Table 4.16: Tank Category-wise Percentage of the Pattern of Land Holding to Total Agricultural Land of the HHs

Category of the Tank	*Own*	*Barga*	*Patta*	*Lease in*	*Lease out*	*Total Agri. Land*
Government	99.4	0.6	0.0	0.0	0.0	100
Personal	97.1	2.3	0.3	1.4	1.1	100
Community	98.0	0.9	0.0	1.2	0.1	100
Total	97.9	1.3	0.1	1.1	0.4	100

Source: Based on Primary (Field) Survey.

4.1.2 Irrigation Characteristics of the Households

Irrigated and Non-irrigated land: We have also collected the data on the distribution irrigated land and non-irrigated land. In Table 4.17, we have given that the percentages of irrigated land and non-irrigated land. The percentages of irrigated land and non-irrigated land are respectively 56 per cent and 44 per cent. Therefore, in most areas there are large amount of lands that are without irrigation.

Table 4.17: Tank Category-wise per cent of Irrigated and Non-irrigated Land to Total Agricultural Land

Category of the Tank	*Irrigated Land*	*Non-Irrigated Land*	*Total Agri. Land*
Government	70	30	100
Personal	53	47	100
Community	54	46	100
Total	56	44	100

Source: Based on Primary (Field) Survey.

Per Capita Irrigated and Non-irrigated Land: In Table 4.18, we have presented that tank category-wise per capita irrigated and non-irrigated land. Here the per capita irrigated and non-irrigated land in Government category are 0.30 acre and 0.13 acre respectively, in Personal category that are 0.22 acre and 0.20 acre respectively and in Community category that are 0.26 acre and 0.22 acre respectively.

Table 4.18: Tank Category-wise Per Capita Irrigated and Non-irrigated Land (In Acre)

Category of the Tank	*Irrigated Land*	*Non-Irrigated Land*	*Total Agri. Land*
Government	0.30	0.13	0.43
Personal	0.22	0.20	0.42
Community	0.26	0.22	0.48
Total	0.25	0.20	0.45

Source: Based on Primary (Field) Survey.

Irrigated and Non-irrigated Land Per Household: In Table 4.19, we have presented that irrigated and non-irrigated land per household. Here the per household irrigated and non-irrigated land in Government category are 1.71 acre and 0.72 acre respectively, in Personal category that are 1.44 acre and 1.30 acre respectively and in Community category that are 1.72 acre and 1.46 acre respectively.

Table 4.19: Tank Category-wise Amount of Irrigated and Non-irrigated Land per Household (In Acre)

Category of the Tank	*Irrigated land*	*Non-Irrigated land*	*Total Agri. Land*
Government	1.71	0.72	2.43
Personal	1.44	1.30	2.74
Community	1.72	1.46	3.19
Total	1.62	1.29	2.91

Source: Based on Primary (Field) Survey.

Percentages of Gross Irrigated Land Distribution: The tank category-wise distribution of gross irrigated land between Aman paddy and other cultivation is shown in Table 4.20. The distribution of gross irrigated land between Aman and non-Aman (other crops) are 70 per cent and 30 per cent respectively for Government category, the percentages for Aman and non-Aman are 77 per cent and 23 per cent respectively for Personal category and that are 68 per cent and 32 per cent respectively for Community category.

Table 4.20: Tank Category-wise Percentages of Gross Irrigated Land Distribution between Aman Paddy and Other Cultivations

Category of the Tank	*Aman Paddy*	*Others*	*Total*
Government	70	30	100
Personal	77	23	100
Community	68	32	100
Total	71	29	100

Source: Based on Primary (Field) Survey.

Average of Gross Irrigated Land Distribution: The tank category-wise distribution of average gross irrigated land between Aman paddy and other cultivation is shown in Table 4.21. The distribution of average gross irrigated land between Aman and non-Aman (other crops) are 1.73 acre and 1.33 acre respectively for Government category, the average for Aman and non-Aman 1.44 acre and 0.99 acre respectively for Personal category and that are 1.74 acre and 1.42 acre respectively for Community category.

Table 4.21: Tank Category-wise Average of Gross Irrigated Land Distribution between Aman Paddy and Others Cultivation

Category of the Tank	*Aman Paddy*	*Others*	*Total*
Government	1.73	1.33	2.47
Personal	1.44	0.99	1.86
Community	1.74	1.42	2.55
Total	1.63	1.29	2.30

Source: Based on Primary (Field) Survey.

Preference Pattern: We had asked the households about the preference pattern (1 to 6) for different types of irrigation. We have calculated the averages of the preferences of the households. Table 4.22 shows that out of 6 options of Tank, Deep, Canal, Shallow, RLI and Others, the preferences of tank for different tank categories are same. The average preference for different sources of the whole sample are as follows: Tank–1st, Deep–4th, Canal–3rd, Shallow–2nd, RLI–5th, Others–6th.

Table 4.22: Tank Category-wise Preference Pattern of Households for Different Sources of Irrigation

Category of the Tank	*Tank*	*Deep*	*Canal*	*Shallow*	*RLI*	*Others*
Government	1st	4th	3rd	2nd	5th	5th
Personal	1st	4th	3rd	2nd	5th	5th
Community	1st	5th	2nd	3rd	4th	5th
Total	1st	4th	3rd	2nd	5th	5th

Source: Based on Primary (Field) Survey.

Irrigated Land Position: The land of the surveyed households may be at different positions of the tank—head, tail, mixed. Mixed means some parts of land are of flow irrigation and some parts are of lift irrigation. The tank category-wise percentages of households with respect to their positions of the tanks are presented in Table 4.23. The percentages for the whole sample are as follows: Head–8 per cent, Tail–66 per cent and Mixed–26 per cent.

Table 4.23: Tank Category-wise per cent of HHs having Irrigated Land Position from the Tank

Category of the Tank	*Head*	*Tail*	*Mixed*	*Total*
Government	13	59	28	100
Personal	7	73	20	100
Community	5	65	30	100
Total	8	66	26	100

Source: Based on Primary (Field) Survey.

Percentages of Irrigated Land: We have collected data on the irrigation by tanks for Aman and Other crops. These have been classified into different categories of tank ownership. Table 4.24 shows that the distributions of irrigated land by tank for Aman paddy are as follows: Government–70.3 per cent, Personal–76.7 per cent and Community–68.4 per cent and for the other crops are as follows: Government–29.7 per cent, Personal–23.3 per cent and Community –31.6 per cent. Therefore, there is not much difference between Government, Personal and Community category in respect of distribution of land between Aman and Boro crops.

Table 4.24: Tank Category-wise Percentages of Irrigated Land from Tank for Aman and Other Crops

Category of the Tank	*Aman Paddy*	*Others*	*Total*
Government	70.3	29.7	100.0
Personal	76.7	23.3	100.0
Community	68.4	31.6	100.0
Total	73.1	26.9	100.0

Source: Based on Primary (Field) Survey.

4.1.3 Willingness to Pay (WTP) for Tank Construction/Renovation

Willingness to pay for tank maintenance is an important parameter of the sustainability of tanks.

Annual WTP for Tank Maintenance: We have collected the data the households' willingness to pay for tank maintenance from our field survey. The average WTP for tank maintenance in the three categories, Government, Personal and Community are Rs. 116, Rs. 173 and Rs. 199 respectively. In Table 4.25 we see that the average willingness to pay of Community category is more compared to the other two categories.

Table 4.25: Tank Category-wise Annual WTP for Tank Maintenance

Category of the Tank	*Average Willingness to pay per year for tank maintenance (Rs.)*
Government	116
Personal	173
Community	199
Total	177

Source: Based on Primary (Field) Survey.

Percentage of the HHs Willing to pay: In Table 4.26, we have presented frequency distribution of households' willingness to pay for different categories. This table shows that out of 533 households 23 per cent households are not willing to pay. The number of households' willingness to pay within Rs. (1-100) is 45 per cent, within Rs. (101-200) is 13 per cent, within (Rs. 201-500) is 15 per cent, within (Rs. 501-1000) is 4 per cent and above Rs. 1000 is 1 per cent. Highest percentage of households (45%) is willing to pay within Rs. (1-100).

Table 4.26: Tank Category-wise Per cent of the HHs Willing to Pay Money by Payment Group

Category of the Tank	*Rs. 0*	*Rs. (1-100)*	*Rs. (101-200)*	*Rs. (201-500)*	*Rs. (501-1000)*	*Above Rs.1000*	*Total*
Government	15	58	20	6	1	0	100
Personal	14	49	15	20	2	0	100
Community	31	38	10	15	4	2	100
Total	23	45	13	15	3	1	100

Source: Based on Primary (Field) Survey.

4.1.4 District-wise Some More Important Information Relating to Household

In this section we report some other district-wise important information calculated from the household data.

Table 4.27: Some Important Information about Household

Items	*Ban-kura*	*Paschim Madiniur*	*Puru-lia*
Average Annual Family Income (Rs. '000)	10.6	3.4	5.1
% of HHs with selling of produced crops	62	58	54
% of HHs with bank a/c	30	25	45
% of HHs supporting tank irrigation	100	100	98
% of HHs reporting that causes of unutilisation of land is the lack of water	99	72	100
% of HHs reporting that water availability in tank is falling	100	100	100
% of HHs supporting the following causes for water availability in tank is falling			
(i) Soil erosion	97	100	100
(ii) Leakage of bunds	36	44	40
(iii) Seepage	66	74	96
(iv) Siltation	97	67	100
% of HHs reporting that irrigation sources other than tanks are used by them	0	17	9
% of HHs reporting members of SHG	81	48	42
% of HHs reporting attending the Gram Samsad meeting	69	66	67
% of HHs taking credit from different sources	31	17	36
% of HHs taking credit from following different sources			
(i) Bank	19	07	30
(ii) Credit Society	48	47	25
(iii) Panchayat	0	6	0
(iv) Mahajan	33	40	28
(v) Gamin Bank	0	0	17
% of HHs having storage facilities	17	4	0
Average distance (mtr.) between tank and cultivated land of the HHs	76.12	122.42	60.75
Average wage rate (Rs.) per day of the households in present	56	68	55

Source: Based on Primary (Field) Survey.

- Migrations of the household's members to other places are reported to be low.
- District-wise percentage of households reporting that they are contracted with input suppliers are very low
- District-wise percentage of households reporting crop failure is also low.
- Land class-wise and crop-wise average cost of cultivation and profit have been calculated.
- Household contribution in tank management is found to be low.

4.2 Cropping Pattern and Irrigation Cost for Different Categories of Tanks

Percentage of Household cultivating different types of crops: In Table 4.28 we have presented tank category-wise percentage of household cultivating different types of crops. Here we see that in all three categories all households are cultivating Aman Paddy. On the other hand the 7.9 per cent households are cultivating Boro Paddy, 22.9 per cent—Wheat, 9.9 per cent—Potato, 0.2 per cent—Groundnut, 6.9 per cent—Vegetable, 23.8 per cent—Mustard and 9.0 per cent Seasame.

Table 4.28: Tank Category-wise Percentage of Household Cultivating Different Types of Crops

Category of the Tank	*Aman Paddy*	*Boro Paddy*	*Wheat*	*Potato*	*Ground Nut*	*Vege-table*	*Mustard*	*Sea-smum*
Government	100.0	11.8	12.9	10.6	0.0	3.5	24.7	12.9
Personal	100.0	6.0	11.5	7.1	0.5	4.4	23.0	9.8
Community	100.0	7.9	34.0	11.7	0.0	9.8	24.2	7.2
Total	100.0	7.9	22.9	9.9	0.2	6.9	23.8	9.0

Source: Based on Primary (Field) Survey.

Average times of Irrigation: We have calculated the numbers of irrigation for different types of crops. We have observed that a time of irrigation of Boro paddy is much higher than the Aman paddy and other crops. Table 4.29 has been prepared on the basis of the response of the households. The average times of irrigation for boro paddy for Government, Personal and Community category are respectively 16, 15 and 14.

Table 4.29: Tank Category-wise Average Times of Irrigation from the Tank

Category of the Tank	Aman Paddy	Boro Paddy	Potato	Mustard	Wheat	Til	G. Nut	Sunflower
Government	2	16	6	3	4	3	—	—
Personal	2	15	7	4	4	2	6	—
Community	2	14	6	3	4	3	—	5
Total	2	15	6	4	4	3	6	5

Source: Based on Primary (Field) Survey.

Irrigation Cost for Different Crops: The costs of cultivation of different crops have been assessed based on household information. The block-wise average irrigation cost for different types of crops is presented in Table 4.30. The average irrigation cost per acre for different types of crops are as follows: Aman Paddy–0.2 thousand, Boro Paddy–2.2 thousand, Wheat–1.4 thousand, Potato–1.7 thousand, Groundnut–0.9 thousand, Vegetable–1.4 thousand, Mustard–1.0 thousand and 0.6 thousand. The irrigation cost per acre is highest for Boro Paddy followed by Potato. So the irrigation cost is high in Boro Paddy and Potato compared to other crops.

Table 4.30: Tank Category-wise Average Irrigation Cost for Different Crops of Households

(Rs. in thousands/acre)

Category of the Tank	Aman Paddy	Boro Paddy	Wheat	Potato	Ground Nut	Vegetable	Mustard	Sesamum
Government	0.3	2.1	1.8	1.4	—	—	1.0	0.7
Personal	0.2	2.0	1.5	1.2	0.9	1.3	1.1	0.5
Community	0.2	2.2	1.3	1.9	—	1.6	1.0	0.7
Total	0.2	2.2	1.4	1.7	0.9	1.4	1.0	0.6

Source: Based on Primary (Field) Survey.

Average Tank Irrigation Cost: As we have already stated the tank irrigation is the major irrigation in our survey areas. We have calculated the tank irrigation cost crop-wise. The grand averages per acre of the three tank categories are as follows (Table 4.31): Aman Paddy–0.2 thousand, Boro Paddy–1.9 thousand, Wheat–1.3 thousand, Potato–1.4 thousand, Vegetable–1.1 thousand, Mustard –1.0 thousand and Sesamum–0.5 thousands. The tank irrigation cost per acre is highest for Boro Paddy followed by others.

Table 4.31: Tank Category-wise per Acre Average Tank Irrigation Cost for Different Crops

(in '000 Rs.)

Category of the Tank	Aman Paddy	Boro Paddy	Wheat	Potato	Vegetable	Mustard	Sesamum
Government	0.3	1.9	1.8	1.0	0.0	1.0	0.6
Personal	0.2	1.9	1.4	1.1	1.3	1.0	0.5
Community	0.2	1.8	1.2	1.7	1.2	0.9	0.6
Total	0.2	1.9	1.3	1.4	1.1	1.0	0.5

Source: Based on Primary (Field) Survey.

Share of Tank Irrigation Cost to Total Irrigation Cost: We have calculated the tank irrigation cost as a percentage of total irrigation cost for different crops for different tank categories. Table 4.32 shows that the percentage of tank irrigation cost in total irrigation cost for different crops are as follows: Aman Paddy–100 per cent, Boro Paddy–86 per cent, Wheat–93 per cent, Potato–87 per cent, Vegetables–81 per cent, Mustard–84 per cent and Sesamum–83 per cent. As we have discussed earlier other crops like Boro, Wheat, Potato uses more complementary sources of irrigation.

Table 4.32: Tank Category-wise Per Acre Share of Tank Irrigation Cost to Total Irrigation Cost

Category of the Tank	Aman Paddy	Boro Paddy	Wheat	Potato	Vegetable	Mustard	Sesamum
Government	100	89	100	77	—	99	85
Personal	99	94	93	92	100	98	84
Community	100	81	91	87	76	90	80
Total	100	86	93	87	81	94	83

Source: Based on Primary (Field) Survey.

Share of Tank Irrigation Cost to Total Production Cost: Land-class-wise tank irrigation cost as a percentage of production cost for different tank category is presented in Table 4.33. The tank irrigation cost as a percentage of production cost for different crops is calculated as follows: Aman Paddy–4 per cent, Boro Paddy–22 per cent, Wheat–22 per cent, Potato–11 per cent, Vegetables–10 per cent, Mustard–18 per cent and Sesamum–19 per cent. It may be noted that tank irrigation cost as percentage of production cost is lower for Aman paddy compared to other crops. This may be due to the fact that other components of production cost is lower for aman crop and other irrigation sources are used for boro crop.

Table 4.33: Tank Category-wise Per Acre Share of Tank Irrigation Cost to Total Production Cost

Category of the Tank	Aman Paddy	Boro Paddy	Wheat	Potato	Vegetable	Mustard	Seasmum
Government	4	22	25	7	—	18	20
Personal	4	23	22	8	21	19	16
Community	4	21	21	13	9	17	20
Total	4	22	22	11	10	18	19

Source: Based on Primary (Field) Survey.

4.3 ANOVA Analysis of Value, Cost and Profit for Different Crops Cultivated by Sample Households in the Tank Command Area for Different Tank Categories

In this sub-section we will compare value, cost and profit for different crops and will check whether there is any significant difference among these for different categories of tanks.

Analysis of Variance (Anova): Anova analysis is used to compare the means of more than two tank categories. This analysis uses the F-statistic, which tests if the means of the groups, formed by independent variable or a combination of independent variables, are significantly different. It is based on the comparison of two estimates of variances— (i) Variance within groups (error variance), (ii) Variances due to differences in group means. F is defined as: F = Variance due to difference between groups/Error Variance.

The null hypothesis is written as H_0 = The average value/acre, cost/acre or profit/acre of a crop are same in all the tank categories.

ANOVA for Aman Paddy: In Table 4.34, we have calculated tank category-wise value, cost and profit per acre for Aman Paddy. Here we see that the value/acre, cost/acre and profit/acre are Rs. 11.3 thousand, Rs. 6.0 thousand, and Rs. 5.2 thousand respectively for the Government category; for the Personal category these are Rs. 10.7 thousand, Rs. 6.0 thousand, and Rs. 4.7 thousand respectively and for the Community category these are Rs. 10.5 thousand, Rs. 6.1 thousand, and Rs. 4.4 thousand respectively.

The output of one-way ANOVA is shown in the following Table. F value of value/acre, cost/acre and profit/acre are 8.78, 1.27 and 12.58 and the corresponding P-value is 0.000176, 0.280969 and 4.56E-06. Therefore, we can safely reject the null hypothesis and conclude that average value/acre, profit/acre of Aman Paddy are not same in the tank categories.

Table 4.34: Tank Category-wise Average Value, Cost and Profit Per Acre for Aman Paddy

(Rs. in '000)

Category of the Tank	*Value (in Rs./acre)*	*Cost (in Rs./acre)*	*Profit (in Rs./acre)*
Government	11.3	6.0	5.2
Personal	10.7	6.0	4.7
Community	10.5	6.1	4.4
Total	10.7	6.0	4.6
ANOVA			
F	8.78696	1.272556	12.58865
P-Value	0.000176	0.280969	4.56E-06

Source: Based on Primary (Field) Survey.

ANOVA for Boro Paddy: In Table 4.35, we have calculated land class-wise value, cost and profit per quintal for Boro Paddy. Here we see that the value/acre, cost/acre and profit/acre are Rs. 13.4 thousand, Rs. 8.7 thousand, and Rs. 4.7 thousand respectively for the Government category; for the Personal category these are Rs. 13.4 thousand, Rs. 8.5 thousand, and Rs. 4.9 thousand respectively and for the Community category these are Rs. 13.8 thousand, Rs. 8.5 thousand, and Rs. 5.3 thousand respectively.

The output of one-way ANOVA is shown in the following Table 4.35. F value of value/acre, cost/acre and profit/acre are 0.26, 0.06 and 0.13 and the corresponding P-value is 0.76919, 0.935209 and 0.871052. Therefore, we can safely reject the null hypothesis and conclude that average value/acre, profit/acre of Aman Paddy are not same in the three districts.

ANOVA for Wheat: In Table 4.36, we have calculated land class-wise value, cost and profit per quintal for Wheat. Here we see that the value/acre, cost/acre and profit/acre are Rs. 9.7 thousand, Rs. 7.4 thousand, and Rs. 2.3 thousand respectively for the Government category; for the Personal category these are Rs. 7.7

Table 4.35: Tank Category-wise Average Value, Cost and Profit Per Acre for Boro Paddy

(Rs. in '000)

Category of the Tank	Value (in Rs./acre)	Cost (in Rs./acre)	Profit (in Rs./acre)
Government	13.4	8.7	4.7
Personal	13.4	8.5	4.9
Community	13.8	8.5	5.3
Total	13.6	8.5	5.0
ANOVA			
F	0.26419	0.0671	0.138544
P-Value	0.76919	0.935209	0.871052

Source: Based on Primary (Field) Survey.

thousand, Rs. 6.1 thousand, and Rs.1.6 thousand respectively and for the Community category these are Rs. 7.3 thousand, Rs. 6.0 thousand, and Rs. 1.4 thousand respectively.

The output of one-way ANOVA is shown in the following Table. F value of value/acre, cost/acre and profit/acre are 12.30, 7.51 and 3.95 and the corresponding P-value is 1.39E-05, 0.000847 and 0.0218. Therefore, we can safely reject the null hypothesis and conclude that average value/acre, cost/acre of wheat are not same in the three tank categories.

Table 4.36: Tank Category-wise Average Value, Cost and Profit Per Acre for Wheat

(Rs. in '000)

Category of the Tank	Value (in Rs./acre)	Cost (in Rs./acre)	Profit (in Rs./acre)
Government	9.7	7.4	2.3
Personal	7.7	6.1	1.6
Community	7.3	5.9	1.4
Total	7.6	6.0	1.5
ANOVA			
F	12.3019	7.511035	3.952596
P-Value	1.39E-05	0.000847	0.0218

Source: Based on Primary (Field) Survey.

ANOVA for Mustard: In Table 4.37, we have calculated land class-wise value, cost and profit per quintal for Mustard. Here we see that the value/acre, cost/acre and profit/acre are Rs. 14.4 thousand,

Rs. 5.6 thousand, and Rs. 8.8 thousand respectively for the Government category; for the Personal category these are Rs. 12.7 thousand, Rs. 5.6 thousand, and Rs. 7.1 thousand respectively and for the Community category these are Rs. 16.0 thousand, Rs. 5.3 thousand, and Rs. 10.7 thousand respectively.

The output of one-way ANOVA is shown in the following Table 4.37. F value of value/acre, cost/acre and profit/acre are 6.78, 0.72 and 9.09 and the corresponding P-value is 0.001597, 0.486755 and 0.000207. Therefore, we can safely reject the null hypothesis and conclude that average value/acre, profit/acre of Mustard are not same in the tank categories.

Table 4.37: Tank Category-wise Average Value, Cost and Profit Per Acre for Mustard

(Rs. in '000)

Category of the Tank	*Value (in Rs./acre)*	*Cost (in Rs./acre)*	*Profit (in Rs./acre)*
Government	14.4	5.6	8.8
Personal	12.7	5.6	7.1
Community	16.0	5.3	10.7
Total	14.6	5.5	9.2
ANOVA			
F	6.785973	0.724191	9.090302
P-Value	0.001597	0.486755	0.000207

Source: Based on Primary (Field) Survey.

ANOVA for Potato: In Table 4.38, we have calculated land class-wise value, cost and profit per quintal for Potato. Here we see that the value/acre, cost/acre and profit/acre are Rs. 28.2 thousand, Rs. 15.4 thousand, and Rs. 12.8 thousand respectively for the Government category; for the Personal category these are Rs. 27.9 thousand, Rs. 14.1 thousand, and Rs. 13.8 thousand respectively and for the Community category these are Rs. 28.8 thousand, Rs. 12.5 thousand, and Rs. 16.3 thousand respectively.

The output of one-way ANOVA is shown in the following Table 4.38. F value of value/acre, cost/acre and profit/acre are 0.02, 1.74 and 1.68 and the corresponding P-value is 0.970545, 0.184511 and 0.196078. Therefore, we can safely reject the null hypothesis and conclude that average value/acre, profit/acre of Mustard are not same in the tank categories.

Table 4.38: Tank Category-wise Average Value, Cost and Profit Per Acre for Potato

(Rs. in '000)

Category of the Tank	*Value (in Rs./acre)*	*Cost (in Rs./acre)*	*Profit (in Rs./acre)*
Government	28.2	15.4	12.8
Personal	27.9	14.1	13.8
Community	28.8	12.5	16.3
Total	28.5	13.3	15.1
ANOVA			
F	0.029916	1.74848	1.683502
P-Value	0.970545	0.184511	0.196078

Source: Based on Primary (Field) Survey.

4.3.1 Input Cost Shares for Different Crops

We have also collected the input costs of production for different crops. In Table 4.39 below we have presented the average input cost shares for different crops for the sample households in the tank command area. As the table shows tank irrigation costs are high for the crops like boro, wheat, potato and mustard.

Table 4.39: Average Input Cost Share for Different Crops for the Sample Farmers

	Lobour	*Bullock*	*Capital*	*Pesti-cides*	*Tank Irri-gation*	*Other Irri-gation*	*Seeds*	*Fertili-sers*	*Total*
Aman	51.99	12.33	5.98	4.01	3.52	0.01	4.45	17.71	100
Boro	36.93	10.58	4.19	5.02	19.84	2.09	4.22	17.13	100
Wheat	28.42	13.72	6.16	1.67	23.29	0.68	7.07	17.16	100
Potato	19.72	9.44	4.79	5.92	16.90	2.68	18.31	22.25	100
Mustard	25.38	14.49	3.76	5.32	19.92	0.74	5.16	20.00	100

4.4 Relation between Farm Size and Profit per Acre

We have analysed the relation between farm size and profit per acre of the households for different categories of tanks. We have estimated by the following regression equation of the households:

$$PA = a + bFS + U_i$$

where PA is the profit per acre (Rs.) = Net Profit/Gross Cropped Area, FS = Farm Size. We have first estimated the regression equation for the Government categories of tanks. The estimated coefficient is negative and statistically significant. This implies that there is statistically significant negative relation between farm size and profitability in this type of tank categories.

Table 4.40: Regression result for Farm size and Profitability for the Government Category of Tanks

ANOVA

	Df	*SS*	*MS*	*F*	*Significance F*
Regression	1	6871958	6871958	4.878876	0.029944
Residual	83	1.17E+08	1408512		
Total	84	1.24E+08			

	Coefficients	Standard Error	*t* Stat	P-value
Intercept	5646.865	208.913	27.02974	6.66E-43
FSG	—118.327	53.57031	—2.20882	0.029944

FSG = Farm Size in Government Category, Dependent variable: Profit per acre Government Category

The relation between farm size and profitability per acre in Personal categories of tanks is tested to be negative but not statistically significant.

Table 4.41: Regression Result for Farm size and Profitability for the Personal Category of Tanks

ANOVA

	df	*SS*	*MS*	*F*	*Significance F*
Regression	1	2257827	2257827	0.798548	0.372715
Residual	181	5.12E+08	2827416		
Total	182	5.14E+08			

	Coefficients	Standard Error	t Stat	P-value
Intercept	4941.367	216.7404	22.79855	4.28E-55
FSP	-51.3143	57.42332	-0.89361	0.372715

FSP = Farm Size in Personal Category, Dependent variable: Profit per acre in Personal Category

The relation between farm size and profitability per acre in Community categories for the sample households in the tank command area is tested to be negative and statistically significant as presented in the following Table 4.42.

Table 4.42: Regression Result for Farm Size and Profitability for the Community Category of Tanks

ANOVA

	df	*SS*	*MS*	*F*	*Significance F*
Regression	1	35086760	35086760	3.819055	0.051732
Residual	263	2.42E+09	9187289		
Total	264	2.45E+09			

	Coefficients	Standard Error	t Stat	P-value
Intercept	5363.082	267.6487	20.03776	7.33E-55
FSC	–96.1605	49.20606	–1.95424	0.051732

FSC = Farm Size in Community Category, Dependent variable: Profit/acre in Community Category

The results on farm size and productivity show that small farms are more efficient for the household under community and government category tanks.

4.5 Willingness to Pay (WTP) Study

4.5.1 Tobit Estimation

We had asked each household about their willingness to pay (WTP) if the new tank/new bund is constructed or tank/bund is renovated and the water supply is ensured for them. The question was open-ended one. We have different responses about their willingness to pay. Some of the respondents are willing to pay zero amount or they avoided to answer the question. From their willingness to pay behaviour, we have tried to measure the Total Economic Value (TEV) of the three districts under consideration.

Firstly, we have used the Tobit regression model to identify the determinants of willingness to pay for improving the Bund condition. Then we have estimated the total willingness to pay. The persons who are not willing to pay or have not expressed anything about their willingness are assumed to have 0 WTP. The negative value of WTP is not observable. The dependent variable is censored at the value 0. Thus in this analysis, the Tobit model can be written as

$$\begin{aligned} WTP &= WTP^*, \text{ if } WTP^* > 0 \\ &= 0, \text{ otherwise} \end{aligned}$$

It should be mentioned that method of estimation procedure depends upon the structure of the bidding game. If the bidding game is closed ended type, the relevant estimation procedure would be logit/ probit. For open-ended question like ours, censored regression like Tobit is more appropriate. We have taken the following model to find the determinants of WTP.

The model is

$$WTP^* = b_1 + b_2\ HEADED + b_3\ FAMEXP + b_4\ TANKQUAL + b_5\ WATERDD$$

HEADED : Education of the head of the household

FAMEXP : Family Expenditure (Rs./month)

TANKQUAL : Score value of the quality of tank as judged by the households out of 100 based on different quality parameters

WATERDD : Water requirement (Acre-mm) calculated from the area cultivated by households for different crops and water requirement for those crops.

We have tested the WTP model with the whole sample of 533 households in three districts. The estimated results are shown in Table 4.43.

Table 4.43: Tobit Estimation Result of WTP

	Coefficient	*Std. Error*	*z-Statistic*	*Prob.*
C	82.49120	145.3455	0.567552	0.5703
HEADED	9.543463	4.420112	2.159100	0.0308
FAMEXP	0.030330	0.013190	2.299378	0.0215
TANKQUAL	–4.867150	2.852038	–1.706551	0.0879
WATERDD	0.038641	0.007326	5.274484	0.0000

Source: Own Estimation.

The above results indicate that the decision of the individuals to contribute or not to contribute for tank maintenance depends significantly upon education of the head of the family (HEADED), family expenditure (FAMEXP), and Water demand for crops cultivated (WATERDD) and negatively with Quality of the tank (TANKQUAL).

4.5.2 Estimation Total Willingness to Pay or Total Economic Value (TEV) of Tanks

Demand for a resource depends upon total economic value (TEV). TEV consists of both use value and non-use value. As water for the tank is used for irrigation, bathing and fishery purposes, it has certainly use value. Contingent valuation method (CVM) has been used for the valuation of use value (*Chandrasekaran 2009, Dutta M. et al. 2005*) though CVM is generally used for valuation of non-use value. We have already stated our sample consists of 533 households. We have collected information about their annual willingness to pay. Total willingness to pay has been calculated by median WTP multiplied by the relevant population. Total willingness to pay reflects the demand for Bund/tank. In the following Table 4.44 we have reported the number of holdings for different size classes for the three districts that we have collected for our survey. There are 13.48 lakh holdings in total in the three districts.

Table 4.44: District-wise Numbers of Holdings for Different Land Classes

	Marginal	*Small*	*Semi-medium*	*Medium*	*Large*	*Total*
Bankura	239365	84960	36870	5827	20	367042
Paschim Medinipur	577469	82262	19001	1478	23	680233
Purulia	213085	63750	22455	2197	27	301514
Total	1029919	230972	78326	9502	70	1348789

Out of 13.48 lakh land holdings, we have tried to estimate the number of households those who are willing to pay on the basis of our sample observation. Out of 533 households surveyed we have observed that 121 households are not willing to pay for the renovation/ reconstruction of Bund. The distribution of 121 households by size classes are presented in the third row of the following Table 4.45. In the next row we have estimated the number of households in the district not willing to pay given the proportion of households not willing to pay in the sample. Then we have estimated the number of households willing to pay in the three districts.

Table 4.45: Estimation of Numbers of Holdings Willing to Pay in the Three Districts

	Marginal	*Small*	*Semi-medium*	*Medium*	*Large*	*Total*
0 WTP (Nos. in sample)	46	28	30	11	6	121
0 WTP (proportion in sample)	0.380	0.231	0.248	0.091	0.050	1.000
0 WTP (Population)	391539	53448	19420	864	3	465274
estimated no of holdings Willing to pay	638380	177524	58906	8638	67	883515

We have collected the data of willingness to pay for all the 533 households in the district. Out of these 121 households have 0 willingness to pay. So the rest of 412 households have some amount of willingness to pay. From these 412 households we have calculated the median willingness to pay for different categories of farmers, which has been reported in the third row of the following Table 4.46. In the last row, we have reported the population WTP for the three districts. The total willingness to pay or TEV is calculated as 6.07 crore.

Table 4.46: Estimation of Total Willingness to Pay in the Land Classes

	Marginal	*Small*	*Semi-medum*	*Medium*	*Large*	
No of holdings Willing to pay	638380	177524	58906	8638	67	
Median WTP (Rs.)	50	100	150	250	1000	
Total WTP (Rs.)	31918977.27	17752393	8835950	2159545	66528.93	60733395

4.6 Factor Analysis: Principal Component Analysis for the Quality of Tanks under Study

Factor analysis is an interdependent technique when primary purpose is to define the underlying structure among variables in the analysis (*Meyers et al. 2006*). Factor analysis enables us for analyzing the structure of the interrelationship among a large number of variables by defining sets of variables that are highly correlated known as factors. We have used principal component analysis for factor analysis. We have taken 13 factors to judge the quality of the tanks. The variables are taken as attribute variables and takes values 1 to 5 as judged by the farmers. We have included the following variables for tank quality:

(1) Adequacy of tank water – ADEWATER

(2) Control of water use – CONTWATER
(3) Water Entry and Exit System - ENTRYEXI
(4) Water Sharing System – WATRSHA
(5) Maintenance of Channels – MAINCH
(6) Maintenance of tanks – MAINTANK
(7) Lining between Channels and Land – LINING
(8) Certainty of getting water – CERTAINT
(9) Role of Panchyat in tank development – PANCHYAT
(10) Suez system – SUEZ
(11) Afforestation – FOREST
(12) Relation among the members – RELATION
(13) Fund availability – FUNDAV
(14) Maintenance Bund of the tank – MAINBUND
(15) Panchayat Tax – PANCHYAT

The scale values for the attributes are as follows: Very Bad–1, Bad–2, Medium–3, Good–4, Very Good–5. In the following Table 4.47, we report the average assessment of tank qualities on the basis of score given by households.

Table 4.47: Average Assessment of Tank Quality Factors

Tank Quality Variables	*Short name*	*Bankura*	*PM*	*Purulia*	*Total*
Adequacy of tank water	ADEWATER	Medium	Medium	Medium	Medium
Control of water use	CONTWAT	Medium	Medium	Medium	Medium
Water entry/exit System	ENTRYEXI	Medium	Medium	Medium	Medium
Water Sharing System	WATRSHA	Medium	Medium	Medium	Medium
Maintenance of Channels	MAINCH	Very Bad	Very Bad	Very Bad	Very Bad
Maintenance of tanks (disltation etc.)	MAINTANK	Very Bad	Bad	Very Bad	Very Bad
Channel/lining between Tank and land	LINING	Very Bad	Bad	Very Bad	Very Bad
Certainty of getting water	CERTAINT	Medium	Bad	Medium	Bad
Role of Panchayat	PANCHYAT	Very Bad	Bad	Very Bad	Very Bad
Maintenance of pond bundh	MAINBUND	Very Bad	Very Bad	Very Bad	Very Bad
Suez System (lock gate)	SUEZ	Very Bad	Very Bad	Very Bad	Very Bad
Forestation	FOREST	Bad	Bad	Very Bad	Bad
Relation: owner–Beneficiary	RELATION	Good	Medium	Medium	Medium
Fund availability	FUNDAV	Very Bad	Very Bad	Very Bad	Very Bad

Source: Based on Primary (Field) Survey.

For factor analysis firstly, we have computed the unrotated factor matrix which contains the factor loading for each variable on each factor. The degree of correspondence between factor and variable is known as loading.

The KMO measure of sampling adequacy is 0.804, which is a rough indicator of how adequate the correlations are for factor analysis. As its value is greater than 0.7, it may be taken as adequate. A significant Barlett's test enables us to reject the null hypothesis of the lack of sufficient correlation among the variables.

KMO and Bartlett's Test

Kaiser-Meyer-Olkin Measure of Sampling Adequacy		.804
Bartlett's Test of Sphericity	Approx. Chi-Square	3234.434
	Df	91
	Sig.	.000

The communalities have presented in the following table. Communalities indicate the weights by which each degree to which each variable is participating or contributing to the components solutions. The initial communalities are 1 because each variable is fully involved in the solution. The extraction communalities indicate the degree to which each variable is captured by the component solution. As extraction communalities are greater than 0.4 for all the variables, there are no variables needed to be removed.

Communalities

	Initial	*Extraction*
ADEWATER	1.000	.600
CONTWAT	1.000	.677
ENTRYEXI	1.000	.550
WATRSHA	1.000	.694
MAINCH	1.000	.795
MAINTANK	1.000	.636
LINING	1.000	.843
CERTAINT	1.000	.606
PANCHYAT	1.000	.808
MAINBUND	1.000	.721
SUEZ	1.000	.673
FOREST	1.000	.507
RELATION	1.000	.440
FUNDAV	1.000	.750

Extraction Method: Principal Component Analysis.

The following Table summarizes the amount of variance accounted for each factor. Each row provides information on a component. The first component accounts for 32.8 per cent variation, followed by second component (15.93%), third component (10.08%), fourth component (7.559%). The unrotated component matrix is shown in the following Table. It shows the correlations between variables and components. Thus variable MAINCH is correlated 0.726 with component –1, –0.387 with component-4. In the component matrix, variables ordered by the correlation with each component.

Eigenvalue indicates the amount of variance of each factor. The eigenvalue of the first factor 4.6 indicates that the first factor has accounted for about 4.6 units of variance.

Total Variance Explained

Compo-nent	*Initial Eigenvalues*			*Extraction Sums of Squared Loadings*			*Rotation Sums of Squared Loadings*		
	Total	*% of Variance*	*Cumu-lative %*	*Total*	*% of Variance*	*Cumul-ative %*	*Total*	*% of Variance*	*Cumul-ative %*
1	4.6	32.9	32.9	4.6	32.9	32.9	3.2	23.2	23.2
2	2.2	15.9	48.8	2.2	15.9	48.8	2.5	18.2	41.4
3	1.4	10.1	58.9	1.4	10.1	58.9	2.0	14.3	55.7
4	1.1	7.6	66.4	1.1	7.6	66.4	1.5	10.7	66.4
5	0.9	6.2	72.6						
6	0.7	5.2	77.9						
7	0.6	4.2	82.1						
8	0.6	4.2	86.3						
9	0.5	3.6	89.8						
10	0.4	3.1	92.9						
11	0.3	2.5	95.4						
12	0.3	2.2	97.6						
13	0.2	1.4	99.0						
14	0.1	1.0	100.0						

Extraction Method: Principal Component Analysis

The scree plot is shown in figure. The scree plot confirms our previous observation that four component best describes our total variance explained in following table.

The coefficient of rotated component matrix is called factor loading. Rotated component matrix is shown in the following table. Variables are ordered by correlations within each component starting with component 1. It is evident from the table that the first four items were found to correlate to the first component. The next four items correlate high with the second component. Considering the nature of the variables, the first component may be called tank maintenance,

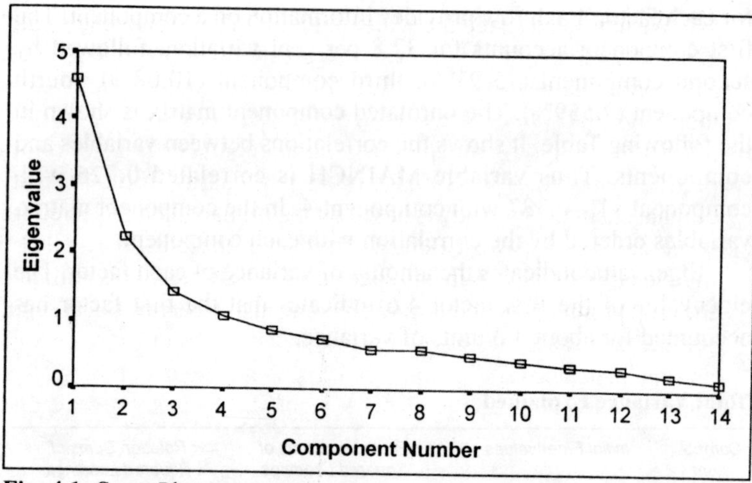

Fig. 4.1: Scree Plot

the second component may be called channel maintenance, the third component may be called water sharing arrangement and the fourth component may be called certainty of getting water.

Rotated Component Matrix

	Component			
	1	2	3	4
PANCHYAT	**.855**	.145	–2.413E-03	–.237
FUNDAV	**.810**	.299	3.983E-02	–5.854E-02
MAINBUND	**.793**	.241	–.126	.137
MAINTANK	**.614**	.501	5.283E-03	–8.494E-02
LINING	.320	**.852**	6.746E-02	.100
MAINCH	.274	**.840**	.103	5.935E-02
SUEZ	.492	**.542**	–.138	.344
RELATION	–.102	**–.423**	.408	.291
WATRSHA	–.118	2.939E-02	**.811**	.144
CONTWAT	.217	–.141	**.774**	.109
ENTRYEXI	–.181	.322	**.638**	–8.429E-02
ADEWATER	–2.843E-02	–6.663E-02	.244	**.732**
CERTAINT	–.436	.130	.109	**.622**
FOREST	.375	.278	–.223	**.489**

Extraction Method: Principal Component Analysis.
Rotation Method: Varimax with Kaiser.
Normalization.
a. Rotation converged in 11 iterations.

We have also tested the scale reliability. The Cronbeench's alpha is calculated as 0.85. As the value is greater than 0.7, it can be used as a seronable test of scale reliability. As the sample size is greater than 500, it is suitable for conducting factor analysis.

Sustainability of irrigation is an important agenda both from the point of view of farmers and farmers particularly in the context of climate change. Our study is based on the primary data collected from some dry zones in West Bengal. The factor analysis reveals that the maintenance of tanks is in bad shape in our study area. This is affecting the crop productivity and farming income in these areas. Local support is essential to ease the budgetary pressure on government for operation and maintenance. We have tried to determine the economic value of tanks through contingent valuation method. The study reveals that if proper institutional mechanism could be developed then farmers are willing to pay for the maintenance of the tanks. The fees will bring an ownership feeling to the farmers so that the co-operation among the farmers for conservation get enhanced.

4.7 Determinants of Productivity and Costs for Households in the Tank Command Area

4.7.1 Value of Farm Output in Tank Command Area

We have analysed the determinants of value of farm output of the households. We have first calculated the value of farm output all the households in our sample. The value of farm output varies across the households. The data for the value of farm output of the households have been tested for various factors. The F test shows the significance of the overall regression. The significant factors are: LOGFERTI (log of fertiliser expenditure), DISTPAKK (Distance from Pakka Road), LOGTANKC (Log of tank irrigation cost), LOGSEED (log of seed expenditure), LOGGCA (log of gross cropped area).

ANOVA for the value of farm output

Model		*Sum of Squares*	*Df*	*Mean Square*	*F*	*Sig.*
1	Regression	416848372710.499	5	83369674542.100	289.857	.000
	Residual	139209691697.197	484	287623329.953		
	Total	556058064407.696	489			

a. Predictors: (Constant), LOGFERTI, DISTPAKK, LOGTANKC, LOGSEED, LOGGCA

b. Dependent Variable: FARMOUTP (Farm Output).

As has been seen from the following table value of farm output is negatively related with the distance of the household from pakka road and fertiliser expenditure. As the distance from pakka road increases there is less marketisation and less production. Tank cost, seed cost and gross cropped areas have positive impact on the value of production of the households.

Table 4.48: Regression Result for the Value of Farm Output

Model		Unstandardized Coefficients B	t	Sig.
1	(Constant)	–15216.665	–.880	.380
	LOGGCA	63811.153	9.575	.000
	LOGTANKC	14062.161	5.347	.000
	LOGSEED	13801.613	3.070	.002
	DISTPAKK	–877.091	–2.957	.003
	LOGFERTI	–15342.199	–2.582	.010

a. Dependent Variable: FARMOUTP

4.7.2 Input Expenditure for Farm Output in Tank Command Area

We have analysed the determinants of input expenditure of the households for crop production. We have first calculated the input expenditure of all the crops produced by households. The input expenditure varies across the households. The input expenditure data has been tested for various factors. The significant factors have been reported in the following table. The F test shows the significance of the overall regression. The significant factors are: MALEWAGE (Male Wage Rate), AGLAND (Agricultural Land), SHAREPOT (Share of Potato), OBC, SHAREVEG (Share of vegetable), SHAREMUS (Share of Mustard), SHAREWHE (Share of Wheat), SHAREBOR (Share of Boro), DOANS (Quality of soil is Loamy), EDUHEAD (Education of Head of the Family). The input expenditure depends negatively with the quality of the soil.

ANOVA

Model		Sum of Squares	Df	Mean Square	F	Sig.
1	Regression	207438643319.516	10	20743864331.952	479.403	.000
	Residual	22587046915.229	522	43270204.818		
	Total	230025690234.745	532			

a. Predictors: (Constant), MALEWAGE, AGLAND, SHAREPOT, OBC, SHAREVEG, SHAREMUS, SHAREWHE, SHAREBOR, DOANS, EDUHEAD

b. Dependent Variable: EXPINP

Table 4.49: Regression Result for Expenditure on Input

Model		Unstandardized Coefficients	t	Sig.
		B		
1	(Constant)	–10425.934	–5.088	.000
	AGLAND	7531.658	63.267	.000
	EDUHEAD	244.374	3.100	.002
	SHAREBOR	258.390	7.146	.000
	SHAREPOT	413.346	8.196	.000
	SHAREVEG	167.680	4.091	.000
	SHAREWHE	304.133	11.164	.000
	SHAREMUS	188.101	5.865	.000
	OBC	408.352	2.356	.019
	DOANS	–1165.287	–1.909	.057
	MALEWAGE	78.745	2.551	.011

a. Dependent Variable: EXPINP

4.7.3 Fertiliser Expenditure (All Crops) in Tank Command Area

We have analysed the determinants of fertilizer expenditure of the households. We have first calculated the fertilizer expenditures all the households in our sample. The fertilizer expenditure on all crops varies across the households. The data for the fertiliser expenditure of the households have been tested for various factors. The F test shows the significance of the overall regression. The significant factors are WEALTHD (wealth dummy), NOFFFARM (No. of off-farm workers), AGEHED (Age of the head), PUMP (whether there is pump), EXTEN (Whether the household gets facilities of extension activities), GALAND (gross agricultural land possessed by the household).

ANOVA

Model		Sum of Squares	Df	Mean Square	F	Sig.
1	Regression	4907541540.104	6	817923590.017	220.705	.000
	Residual	1949330711.689	526	3705951.923		
	Total	6856872251.793	532			

a. Predictors: (Constant), WEALTHD, NOFFFARM, AGEHED, PUMP, EXTEN, GALAND

b. Dependent Variable: VAL. FAT (Fertiliser expenditure for all crops)

The factors with significance levels have been reported in the following Table. It should be noted that fertilizer expenditure is negatively related with EXTEN and NOFFFARM. The households who get the facility of extension activities have lower fertilizer expenditure.

Table 4.50: Regression Results for Fertilizer Expenditure

	Model	*Unstandardized Coefficients*	*t*	*Sig.*
		B		
1	(Constant)	8.264	.029	.977
	GALAND	1235.500	30.978	.000
	PUMP	428.968	1.985	.048
	EXTEN	–531.297	–1.603	.109
	AGEHED	6.455	1.728	.085
	NOFFFARM	–120.527	–1.734	.083
	WEALTHD	924.881	5.730	.000

a. Dependent Variable: VAL. FAT (Fertizer Expenditure).

4.7.4 Productivity in Tank Command Area

4.7.4.1 Determinants of Yield of Aman in Tank Command Area

Aman is the most important crop of all the households. We have analysed the determinants of yield of Aman of the households. We have first calculated the yield of Aman of all the households in our sample. The yield of Aman varies across the households. The data for the yields of Aman have been tested for various factors. The significant factors have been reported in the following Table. The F test shows the significance of the overall regression. The significant factors are SUNFLD (Sunflower Dummy), LOGK (Log Capital), SESAMUMD (Sesame Dummy), BOROD (Boro Dummy), MUSTARDD (Mustard Dummy), CROPSHOC (Crop failure), WHEATD (Wheat Dummy). POTATOD (Potato Dummy), LOGL (Log Labour), LOGLAND

ANOVA

	Model	*Sum of Squares*	*Df*	*Mean Square*	*F*	*Sig.*
1	Regression	441.759	10	44.176	11.744	.000
	Residual	1944.720	517	3.762		
	Total	2386.479	527			

a. Predictors: (Constant), SUNFLD, LOGK, SEASMUMD, BOROD, MUSTARDD, CROPSHOC, WHEATD, POTATOD, LOGL, LOGLAND
b. Dependent Variable: YIELDAMA (Yield of Aman)

Table 4.51: Regression Results for Yield of Aman Paddy

	Model	*Unstandardized Coefficients*	*T*	*Sig.*
		B		
1	(Constant)	–1.885	–.493	.622
	LOGLAND	–6.359	–5.504	.000
	LOGK	1.274	4.318	.000
	LOGL	4.005	3.717	.000
	CROPSHOC	1.361	3.285	.001
	BOROD	.812	2.552	.011
	WHEATD	–.884	–3.991	:000
	POTATOD	–.642	–1.972	.049
	MUSTARDD	.568	2.700	.007
	SEASMUMD	1.237	4.175	.000
	SUNFLD	3.150	3.179	.002

a. Dependent Variable: YIELDAMAN (Yield of Aman Paddy)

4.7.4.2 Productivity of Boro in Tank Command Area

We have analysed the determinants of yield of Boro of the households. We have first calculated the yield of Boro of all the households in our sample. The yield of Boro varies across the households. The data for the yields of Boro have been tested for various factors. The significant factors have been reported in the following table. The F-test shows the significance of the overall regression. The significant factors are DOANSD (Quality of soil is loamy), LOGLAND, EDUHEAD (Education of Head of the Family), AGEHEAD (Age of Head of the Family), VEGD (Vegetable Dummy), LOGK (Log Capital). It should be noted that yield of boro is negatively related the land holding. The smaller landhoding has higher productivity of boro.

ANOVA

	Model	*Sum of Squares*	*Df*	*Mean Square*	*F*	*Sig.*
1	Regression	177.069	6	29.511	9.531	.000
	Residual	89.792	29	3.096		
	Total	266.861	35			

a. Predictors: (Constant), DOANSD, LOGLAND, EDUHEAD, AGEHEAD, VEGD, LOGK

b. Dependent Variable: YIELDBOR (Yield of Boro)

Table 4.52: Estimated Regression Results of Productivity

Model		Unstandardized Coefficients B	t	Sig.
1	(Constant)	–3.297	–.612	.545
	LOGLAND	–1.781	–1.651	.110
	LOGK	2.962	4.262	.000
	EDUHEAD	1.592	1.927	.064
	AGEHEAD	6.749	2.383	.024
	VEGD	2.882	2.507	.018
	DOANSD	3.229	5.257	.000

a. Dependent Variable: YIELDBOR (Yield of Boro Paddy)

4.7.5 Analysis of Agricultural Labour Demand in Tank Command Area

We have analysed the determinants of agricultural labour demand of the households for crop production. We have first calculated the total labour demand of all the crops produced by households. The total labour demand varies across the households. The labour demand data has been tested for various factors. The significant factors have been reported in the following table. The F test shows the significance of the overall regression. The significant positive factors are: MWAGE (Male Wage Rate), MALEADUL (Male Adult), SHAREPOT (Share of Potato), SHARESUN (Share of Sunflower), SHAREMAS (Share of Mustard), EDUHEAD (Education of Head of the Family), SHAREWHT (Share of Wheat), SHAREVEG (Share of Vegetable), DOANSD (Quality of soil is loamy), AGLAND (Agricultural Land), WEALTHD (Wealth Dummy), FEMADULT (Female Adult).

The labour demand depends negatively with the wealth dummy, education of the head of the household and male wage.

ANOVA

Model		Sum of Squares	Df	Mean Square	F	Sig.
1	Regression	16694193.872	12	1391182.823	20.628	.000
	Residual	35070172.451	520	67442.639		
	Total	51764366.323	532			

a. Predictors: (Constant), MWAGE, MALEADUL, SHAREPOT, SHARESUN, SHAREMAS, EDUHEAD, SHAREWHT, SHAREVEG, DOANSD, AGLAND, WEALTHD, FEMADULT

b. Dependent Variable: AGLDD

Table 4.53: Regression Results for Labour Demand

	Model	*Unstandardized Coefficients*	*t*	*Sig.*
		B		
1	(Constant)	340.787	4.339	.000
	AGLAND	27.991	5.167	.000
	SHAREPOT	4.766	2.389	.017
	SHAREVEG	9.227	5.556	.000
	SHAREWHT	2.905	2.702	.007
	SHAREMAS	4.952	3.941	.000
	SHARESUN	22.622	1.998	.046
	WEALTHD	–86.454	–3.487	.001
	DOANSD	45.444	1.874	.062
	EDUHEAD	–15.990	–5.353	.000
	FEMADULT	33.423	2.701	.007
	MALEADUL	41.225	3.852	.000
	MWAGE	–3.929	–3.271	.001

a Dependent Variable: AGLDD

4.8 Land Utilisation of Households in the Tank Command Area: Households Proportion of Boro and Other Crops Cultivation in Gross Cropped Area in Tank Command Area

The performance of the tanks can also be judged whether the second crop is cultivated by the households in the tank command area. The main crop in all the tank command area is Aman crop which is rainfed crop. The households cultivate the second crop if the irrigation water is available. We used the Tobit regression model (Hill Griffiths Lim 2008) to search for the significant factors affecting the probability of growing other crops besides aman. The results show that qualities of the tank as judged by the households, agricultural land holding have significant positive impact on the dependent variable. Here the dependent variable is proportion of gross cropped area used to grow crops other than aman paddy for households.

Household: Dependent variable proportion of other crops (OTHERPROP)

Table 4.54: Tobit Estimation Results for Land Utilisation

Variable	*Coefficient*	*Std. Error*	*z-Statistic*	*Prob.*
C	–0.589547	0.167541	–3.518817	0.0004
EDUHEAD	–0.008060	0.005097	–1.581440	0.1138
QUALITY	0.017783	0.004407	4.034846	0.0001
ALAND	0.014558	0.008421	1.728732	0.0839

Summary of the Chapter

We have analysed the WTP behaviour of the households for the tank renovation/reconstruction with Tobit Analysis. The estimated results show that the significant positive factors for WTP are education of the head of the household and agricultural land of the household. We have also estimated that the total value of Willingness to Pay for tank renovation for the whole population is 6.07 crore in three districts under our study. We have used factor analysis (Principal Component Analysis) for tank quality assessments. Tank quality has been judged on 14 variables. The households perceptions on each variable has been done on 1-5 scale. The results of PCA identify 4 components which identifies 66.4 per cent of the total variation in tank quality variables. The main component has been identified as the tank maintenance affecting the tank quality. The performance of the tanks can be judged whether the second crop is cultivated by the households in the tank command area. Tobit regression analysis has been done to find the determinants of cropping intensity.

Chapter 5

Some Innovative Case Studies on Tank Irrigation

In the previous Chapter, i.e. Chapter 4, we discussed characteristics of households in tank command area, tank quality analysis, willingness to pay behaviour of the households, determinants of productivity, cost and profit for different crops in tank command area. In this chapter we will present two case studies on tank irrigation which are innovative in nature. The cause of presenting them in separate chapter lies in their innovativeness and sustainability. The *happa* model provides an alternative ideal institutional pattern for the dry zones. The first one is on *happa* model and the second on is one watershed management.

5.1 Tank Irrigation with *Happa* Model

5.1.1 Context

There is a specific geographical concentration of backwardness and poverty in some drought-prone areas like western and southern Bankura. These regions are affected by continuous degradation of natural resources, severe erosion, depletion of ground-water reserves, low productivity, low wage rate etc. Any livelihood strategy for improving the living standard of these regions should encompass the programme of arresting the process of degradation of natural resources, restoring ecological balance and, at the same time directly address poverty. The major causes of poverty in these regions are as follows:

(i) Poor farming system support services,
(ii) Lack of efficient system to invest mainstream government fund for land husbandry and
(iii) Lack of farming technical know-how.

Water is a central issue for development in the rainfed dry zones. The rainwater-harvesting tank can play a very vital role in conservation of water resource. The problem with large tank irrigation structures in India is that these are not well managed. The experiment with the formation of water users' association is not satisfactory at all in the state. Some innovative experiments are going on in different parts of India in the irrigation sector. One such experiment in West Bengal is *happa*. Where a small tank called *happa* is being excavated in the private land of the farmer wherefrom the farmer can irrigate his own agricultural land and the tank is managed by the farmer himself. A *happa* is a mud-excavated rain water harvesting structure and does not have any cement work or stone revetment. The sides of a *happa* are stepped with slope of 1:1 such that both livestock and human can access the water of *happa* easily. A *happa* is constructed by the side of agricultural field of a farmer with average length of 45 ft, breadth of 50 ft and depth of 12 ft. The total earth extraction of this *happa* is 17,360 cubic feet which requires 299 mandays. With existing NREGS wage rate of Rs. 100/day the average construction cost with the above specifications is about Rs. 29,900. The average command area of a *happa* is about 0.6 acre. The model is also called 5 per cent model because it occupies 5 per cent of the area of agricultural plot of the farmer. The construction cost of the *happa* is presently being financed from NREGS and all the operational expenditure is being incurred by the farmer for maintaining the *happa*. This model has become successful in some dry zones of West Bengal. It may be mentioned that there are two major cropping seasons in India, namely, Kharif and Rabi. The Kharif season is during the south-west monsoon (July-October). During this season, agricultural activities take place both in rain-fed areas and irrigated areas. The Rabi season is during the winter months (October to June), when agricultural activities take place only in the irrigated areas. Khariff crop includes Aman paddy, maize, pulses etc. Rabi crop includes wheat, barley, oilseeds etc. Construction of water harvesting structures like *happa* have created a strong impact on their livelihood through generation of additional incomes in some dry areas because of the following reasons: (i) Farmers could provide life saving irrigation to paddy crop during this khariff season, (ii) They could grow vegetables around the bund of *happas* etc. It should be pointed out that in most of the dry zones the cropping intensity is poor. One extra crop will have perceptible impact on their standard of living.

5.1.2 Study Area

Our study area belongs to the Biradihi village in Musiaraha G.P. in Hirbandh block which is situated in the western part of Bankura district, West Bengal. The survey was conducted during 2010. In the district Human Development Report of Bankura, Hirbandh has been ranked last out of 22 blocks in the district of Bankura. About 54 per cent of households in the block live below poverty line. There is specific geographical concentration of backwardness in these regions of the district. Only 30 per cent of the agricultural land is irrigated in the block (*Government of West Bengal, District Statistical Handbook 2007*). The per capita annual food grain availability in the block is 230 kg. whereas the requirement is 365 kg. The backwardness can be explained through the lack of access to natural resources like water. The land types in these districts are of three categories: (i) Fallow Uplands (called *Tarh* land): These are at the top of the terrain with very thin topsoil and very low water-holding capacity. (ii) Medium uplands (called *Baid* land): In these types of lands soils are sandy and sandy loam and shallow with low organic matter and low moisture holding capacity. (iii) Low Lands (called *Kanali/Sol* land): These lands are more loamy than *baid* and are most advantageously located in terms of water availability and these lands get additional water from seepage from upper catchment. In these regions 50 per cent-60 per cent land is medium upland, 20 per cent-30 per cent is up land and 30 per cent is low land. The water holding capacity of the barren upland is very low. There is high need for irrigation water in tar and *baid* lands as these possess inferior soil and low moisture. Season-wise paddy is classified into three types—Aus, Aman and Boro. Aman is the main paddy here which is grown in the Khariff season. In the state, September is flowering stage of aman paddy. If there is dry spell, then production of paddy in medium upland is badly affected. The moisture conservation of the soil is very important.

As already mentioned we have selected one village, namely Biradihi for our primary survey. Area of the village is 289.9 ha with total households 106 and the total population of 600. The land use pattern of the village is as follows: Forest–115.3 ha, cultivable waste–54.6 ha, not available for cultivation–48.7 ha, Irrigated land– 20 ha, unirrigated land–61.20 he (*GoI 2001*). So the major part of the agricultural land is unirrigated. Our selected village is in the *baid* land, i.e. medium upland. Agriculture is the main source of income in the village. There is no tube well in the village presently. The only other source of irrigation is *jorh* (*a water harvesting canal which is a common property of the village*). These regions are hotter compared to other regions in West Bengal. The temperature gradient reaches about 45°C in the summer. Though average annual rainfall in the region is about 1400 mm per annum, there is huge run off because of rocky soil and terrain conditions. There is good scope of enhancing irrigation if thus huge runoff is systematically tapped.

5.1.3 Household Characteristics

We have selected 20 households having *Happa* randomly from our sample village. Out of the 20 households selected our calculation reveals the following characteristics of the households:

(i) The average family size of the sample households is calculated as 5.5 and the percentage of male members is 55 per cent.

(ii) The average educational class attained by the head of the household is 3.25 and 42 per cent of the aggregate sample members (excluding child) is illiterate.

(iii) Caste: 18 households are of scheduled caste (SC) category and 2 households are of Scheduled Tribe (ST) category. So all the households belong to backward castes.

(iv) Poverty: 13 households are of BPL (Below Poverty Line) category. Poverty line in India is taken as monthly per capita expenditure of Rs. 356.30 for the rural sector.

(v) Occupation pattern: Only two families have members who are employed in the service sector. The average employment generated per family is calculated as 485 mandays including family labour employed for own agricultural land. The average employment patterns are as follows: Agriculture–74 per cent, non-agriculture–5 per cent, service–8 per cent, NREGS–14 per cent.

(vi) 19 households possess NREGS job cards. Average employment from NREGS per family per year is 59.

(vii) Dependency burden (the percentage of people below the age of 18 years and above 24 years) of the households is calculated as 45 per cent.

(viii) Number of households with different assets is as follows: Mobile phone–5, TV–2, Cow–13, Pumping machine–11.

(ix) Average landholding per family for different categories of land is presented in Table 5.1. Average agricultural land holding is calculated as 0.88 acres per family of which irrigated land from *happa* is 0.35 acres.

(x) Group formation under SGSY scheme is not so good here. Nine households have been reported to belong to Self Help Group.

(xi) There is also the problem of rural indebtedness. 15 households have been observed to take loans from money-lenders. The average loan taken per family is calculated as Rs. 7,250.

(xii) The average expenditure per month per family has been calculated as Rs. 4377. The percentage of expenditure on major items on an average has been calculated as follows: Rice–31 per cent, pulse–4 per cent spices–14 per cent, fish and meat–8 per cent, fruit–2 per cent, medicine–13 per cent, education–21 per cent, cloth–6 per cent, others–2 per cent.

(xiii) The average water level calculated from the household responses is 39.5 ft in the summer season and in the rainy season it is 8.75 ft.

(xiv) Aman paddy is the main crop grown by the households. 16 households have reported that there is Aman crop failure in the current year because of poor rainfall.

(xv) 14 households use other sources of irrigation (mainly *jorh*) except *happa*.

(xvi) We have seen that aman paddy in the study area is generally grown during June to November; vegetables are grown during July to November. In other seasons limited amount of potato, wheat and mustard is grown because of the lack of irrigation facility. The average irrigation numbers for different crops are calculated as follows: Aman paddy–3-5, Vegetables–10-15, wheat–3-4 times.

All the households have given their opinion that water is a great problem in the area and water scarcity is becoming acute in the recent years.

5.1.4 Strategy for Implementation

Any livelihood development strategy in the dry zones should focus

Table 5.1: Average Land Holding per Family for the Sample Households

Land Type	*Amount of land holding (acre)*
Total Land	1.14
Total Agricultural Land	0.88
Total Non-Agricultural Land	0.26
Total Irrigated Land	0.70
Total Non-Irrigated Land	0.18
Irrigated land from *Happa*	0.35
Irrigated land from sources other than *happa*	0.35

Source: Own Estimation from the Primary Survey Data.

on water as the central issue. *In situ* conservation of soil and water along with checking the surface run-off, harvesting of rain-water on the surface, economizing the use of ground-water, rejuvenation of sub-surface water, livelihoods planning through participatory approach at village/hamlet level (based on micro-watershed level flow) have been considered to be at the core of the strategy. Working together with government departments to strengthen Farming system support services influencing local governments to invest on Integrated Natural Resource Management (INRM) based livelihood activities for directly addressing poverty is the urgent need for promoting livelihoods in these areas. Professional Assistance for Development Action (PRADAN) is an organization (NGO)–promoting livelihood in the backward regions. Under a new initiative by the Planning Commission PRADAN was selected as one of the technical resource agencies to facilitate district level planning in Purulia and Bankura districts of West Bengal. Implementing agency is Block level local body named Panchayat Samiti. PRADAN is acting as Project facilitating agency. PRADAN is working in two fronts—orienting the villagers in water conservation and providing the technical support in implementing the project. PRADAN work very closely with PRI system at Panchayat level. The cost of this support is provided through project management cost as allotted in the project. PRADAN has adopted different strategies for addressing poverty like organising women in Small Self Help Group (SHG), helping them in livelihood planning, enabling them to approach local body, administration and bank for getting fund and loan for implementation of the livelihood programme.

As in other places of West Bengal Panchayati Raj Institution (PRI) plays a key developmental role in Bankura also. Panchaayati

Raj is a system of governance in which gram panchayats are the basic units of administration. It has 3 levels: village (called Panchayat), block (Panchayat Samiti) and district (Zilla Parishad). Gram Samsad (Village Council), consisting of 1-2 villages, is the lowest level where village level plans are made. Elected members from Gram Samsad constitute Gram Panchayat (GP). These elected representatives, called Gram Panchayat members, are accountable for preparation and implementation of Annual Plan for the entire Gram Panchayat area (consisting of 12-15 villages). PRIs have different kinds of funds in the form of NREGS to finance the largely labour intensive activities leading to Integrated Natural Resource Management (INRM). The Village level INRM includes the following steps: (i) social mobilization and vision, (ii) delineation of the ridge line and drainage, (iii) mapping the resource, (iv) mapping the land ownership in each, (v) wealth ranking, (vi) preparing land use map, (vii) problem mapping and generating options for remedial, (viii) checking whether all families are adequately addressed, (ix) prioritization and action plan preparation, and (x) proposal preparation.

As revealed from the primary survey the steps followed are given below:

- Mobilising community and grooming of a pool of local Resource Persons (LRPs) through training on INRM plan and implementation. LRPs are selected by Gram Unnayan Committed.
- LRPs prepare village level INRM plan involving all the households and collect the application forms for *happa*.
- LRPs place the plan in the Gram Samsad (GS) meeting/VDC (Village Development Council) and with recommendation of GS/VDC the plan is sent to GP.
- GP issues work order to LRP after getting sanction from block.
- LRP supervises the work and prepares the muster roll of the workers and the Payment is made to the workers with the advice of SAE/Nirman Sahayak/GRS.

It may be mentioned that in most of the areas self help groups are being involved in implementing the programme. In our study area this is not the case because self help groups are not so strong here.

5.1.5 Impacts of the Irrigation Experiment with Happa

The environmental and economic impacts of *Happa* in the regions are reported to be very encouraging. Changes in the irrigated area have always positive impact on rural livelihood, particularly in these reasons where the opportunities for alternative livelihood are very little. Below we mention different kinds of benefits accrued from the programme in the study area.

Economic benefits include the following:

- 100 per cent households surveyed have reported that the yield and cropping intensity of land has increased because of the construction of *happa*. Farmers could provide life saving irrigation to paddy crop during this Kharif season. This has resulted in the improvement of yield. It has been reported that the yield of Aman paddy has increased from 3.5 tonnes to 4.5 tonnes per acre in poor rainfall year.
- They could grow vegetables around the bund of *happa*s and diversify their cropping pattern. Our analysis reveals that cropping intensity has improved from 93 per cent to 102 per cent for the sample farmers in spite of the drought year.
- There is opportunity of generation of wage employment during the construction of *Happa*. The average number of persondays generated for construction of one *happa* is about 300.
- Involvement of local people in the planning and implementation of programme has led to developing more ownership of the programme.
- Out of 20 households surveyed 13 households have done fishery in *happa*. The annual average fishery income per *happa* is calculated as Rs. 1,152.
- Land value of land has increased because of the irrigation facility through *happa*. From the household data we have calculated that the price of the irrigated land is Rs. 1,25,250 per acre and the price of the unirrigated land is Rs. 87,450.
- The *happas* are also meeting the water needs of livestock.

In the following Table 5.2 we have presented the average productivity of different crops for the sample farmers. We see that after construction of *happa*, average productivity has improved for all the crops. This has happened despite the fact that the rainfall is poor compared to the normal this year.

Table 5.2: Average Productivity (Tonnes/hector): Before and After Construction of *Happa*

Crops	*Before Construction Productivity*	*After Construction Productivity*	*Incremental Productivity*
Aus Paddy	4.00	4.45	0.44
Aman Paddy	3.58	4.58	1.00
Potato	6.67	7.46	0.79
Wheat	–	2.04	–
Vegetable	6.34	8.46	2.12
Mustard	–	0.86	–

Source: Own Estimation from the Primary Data.

We have prepared the following Table 5.3 for the aggregate of 20 farmer households that we have surveyed. In Table 5.2 we have presented crop-wise increase in gross cropped area after the construction of *happa*. In West Bengal Paddy is the main production. Paddy is grown in three seasons – Aus, Aman and Boro. As the rainfall was very poor in the surveyed year, paddy production was hampered very much. But the farmers shifted to different types of crops like vegetable (mainly cabbage) with the *happa* water. In Table 5.3 we have presented the crop-wise area before and after construction of *happa*.

Table 5.3: Cropping Pattern: Before and After Construction of *happa* for the Aggregate Sample Households

Crops	*Before Construction*		*After Construction*		*Incremental Total Area*
	Total Area (Acre)	*Nos. of Family*	*Total Area (Acre)*	*Nos. of Family*	
Aus	1.16	04	0.25	02	–0.91
Aman	10.97	13	7.84	08	–3.14
Potato	0.69	06	0.92	07	0.23
Wheat	0.00	00	0.25	02	0.25
Vegetable	3.22	09	8.31	19	5.09
Mustard	0.00	00	0.63	03	0.63
Others	0.33	01	0.00	00	–0.33
Total	16.37		18.19		1.82

Source: Estimation Based on Primary Survey.

5.1.6 Ecological Benefits

As we have already stated the sample *happas* have been constructed within 2009 and 2010. So the full impacts of ecological benefits are yet to be suggested. The soil and moisture conservation of the watershed area has improved in the village. Because of the enhanced moisture retention microbial activities and biomass deposition have increased. As a result local micro environment has improved. 80 per cent of the households surveyed have given their opinion that construction of *happa* has checked soil erosion and run-off. A major share of the households (70%) has opined improvement in land quality. According to the villagers the colour of the soil has changed from red to yellowish and the soil has become more loamy. More herbs and shrubs are now found than before. 25 per cent households have said that water table has improved. All the households are also in favour of NGO involvement in water management.

5.1.7 Social Benefits

Households use water from *happa* for different purposes like bathing, washing clothes, cleaning of utensils. Before the construction of *happa* 45 households had to use only one dug well and one tube well for the purpose of bathing and drinking water. The tube-well is not operating recently. The distress migration in search of work has also been checked. In our survey results we see that number of migration for the sample households has fallen from 21 persons to 15 persons. The process of *happa* construction has also lead to institutional development in terms of greater participation of poor and marginalized farmers, improvement in the relationship between different stakeholders like the farmers, traditional institutions like Panchayat and bureaucracy like block-level authorities.

5.1.8 Economic Viability of Happa: Cost Benefit Analysis

The purpose of this analysis is to have an idea about economic viability of *happa* for our sample which belongs to drought prone region. For this purpose, we have compared the costs of constructing *happa* and annual maintenance costs with the annual benefits generated from *Happa*. As we have already mentioned we have selected 20 *happas* for this analysis. The programme like this obviously generates social and environmental benefits other than economic benefits. We have avoided environmental and social benefits because of the complexity of estimation of these types of benefits. Here we have concentrated only on economic benefits in terms of incremental income from crop production. If the environmental and social benefits are added to the economic benefits, the incremental benefits would be much higher. For the long run viability of the programme we have to see whether the programme generates enough benefits to outweigh costs.

The construction cost of *Happa* is incurred for one time only. Annual costs of tank restoration are assumed to be Rs. 500 per year for the analysis. The incremental returns have been calculated by net profit from increased production from *happa*. The main crops grown in the command area are paddy and vegetables. We have calculated Net Present Value (NPV), Benefit Cost Ratio (BCR) and the Internal Rate of Return (IRR) for CBA using the standard cost benefit analysis technique (*Reddy 2009*). We have assumed that lifetime of the *happa* as 10 years and the discount rate as 15 per cent which is taken as the long term lending rate. All the 20 *happas* in our sample have been constructed within the period 2008 and 2010. The year-wise numbers of *happa*s constructed are as follows: 2008-1, 2009-24 and 2010-1. For our sample the average length of the *happa* is 52 ft with maximum 60 ft and minimum 40ft, the average width of the *happa* is 36 ft with maximum 50 ft and minimum 30 ft and the average depth of the *happa* is 11 ft with maximum 12 ft and minimum 10ft. The average construction cost of the *happa* is Rs. 25,260 with the maximum being Rs. 33,600 and the minimum being Rs. 9,400. The average incremental profit per *happa* has been calculated as Rs. 11,241. Cost benefit analysis results are presented in Table 5.4. We have prepared this table by averaging data of the 20 sample happas under our consideration. The results reveal hat there is much economic justification for the construction of *happa*. Present Value of benefits of the incremental returns assumed to be accrued for the future 10 years is calculated as Rs. 64,682 and the Net Present Value (NPV) is calculated as Rs. 37,036. The Internal Rate of Return (IRR) is about 75.7 per cent which is much higher than the market rate of interest.

Table 5.4: Cost Benefit Analysis Results for a *happa*

	Value (Rs.)
Present Value of Benefit	64,682
Present Value of Cost	27,646
Net Present Value	37,036
Benefit Cost Ratio	2.3
IRR	75.71

Source: Own Estimation from the Primary Survey Data.

The performance of *happa* depends very much on the rainfall. The current year in which we have done our survey has been declared as a drought in the district—the annual rainfall has been only 600 mm where the normal is about 1400 mm. Though this analysis has been done for 20 households, the results will not differ very much if we include more households.

5.1.9 Scaling Up

Scaling-up means enhanced geographical cover. It could also be interpreted as the expansion of number of beneficiaries. Enhanced human resource development at local level is the key input of the viability and the scaling up of technology. The programme has strong potential as judged by sustainability indicators for the farmers like increasing market access, employment opportunities and more control over water resource. Farmers' beliefs and practices can be changed if a campaign is systematically planned and implemented. The key ingredient in participatory research is listening and understanding each other among the stakeholders. As most of the agricultural area is single cropped, there is huge scope of scaling up of this experiment in this area. As the reports come, there is huge demand for *happa* in these regions. Initially the farmers' interest was very low. But the success of the some *happas* has acted as impetus to other farmers to have *happa*. In fact, some households have constructed more than one *happa*—in our sample 20 households have constructed 26 *happa*s in total. Presently the *happa* programme is going on in 3 G.P.s out of 5 G.P.s in Hirbandh block and the progrmme had started from 2008. In Hirbandh block the year-wise numbers of *happas* constructed are as follows: 2008-09: 40, 2009-10: 950 and 2010-11: 1200. It has been reported that about 2,000 *happa*s can be constructed in a single G.P. It has been revealed in the survey that some people prefer *Indara* (dug well) to *happa*. But on sustainability ground it is less preferred because it is based on ground water instead of surface water harvesting. Also the cost of *indara* is high—with the same cost of one *indara* construction, 3 families could be provided with one *happa* each. The major hindrances of upscaling the programme in these regions are as follows: (i) Lack of awareness about the scheme, (ii) Lack of efficient system to invest mainstream government fund for land husbandry, and (iii) Lack of political will.

5.1.10 Conclusion

The small irrigation programme like *happa* has strong impact on the livelihood security of the rural people. It has improved the productivity, intensity and diversity of crops. The diversification of production of farmers from a single khariff crop of paddy into other crops like vegetable production and fishery has reduced their vulnerability to climate shock they faced earlier before the construction of *happa*. The success and up scaling of the programme depends very much on the system of planning, application, execution, monitoring and fund-flow. As we have noted from the field survey, there should be more emphasis on crop diversification. Paddy cultivation is more risky compared to crops like maize and vegetables because in case of low rainfall paddy crop may suffer heavily. Our survey results reveal that 100 per cent households have given their opinion that they fail to utilize the land because of the lack of water. There is huge demand for irrigation facility in these areas. Government should take more pro-active role in upscaling the experiment according to the needs. To make it success technological interventions are required in terms of new production techniques like SRI cultivation and new irrigation techniques like drip and sprinkler irrigation techniques for conserving water and also organic farming method. It must be pointed out that average annual rainfall in the district of Bankura, West Bengal is about 1400 mm. So, there is huge scope of enhancing irrigation if the runoff water is systematically and properly tapped. Another big advantage of this model is that number of beneficiaries per unit expenditure spent is much higher in *happa* than the bigger irrigation model. As the *happa* is private property of the individual farmer, farmers have the incentive to maintain the structure and being low cost it is also affordable to them. The strong feature of the *happa* model is that it is both replicable and sustainable. There is enough scope of uplifting of the livelihood of the marginalized sections of the rural community through this programme. More innovations are required how this programme can be integrated with other watershed activities so that water can be more efficiently utilized. Also there is a need to think how small farmers and farmers with scattered land and also landless farmers can avail the benefits of the progrmme. There is also high need of the political will to support the programme by creating awareness of the programme and also involving more women in the decision making process. There should be more emphasis on capacity building at the local level and the development of institutional arrangement. The success and upscaling of the programme depends very much on the system of planning, application, execution, monitoring and fund-flow.

5.2 Watershed Management—A Case Study in Bankura

Watershed management generally entails development of community based organization entrusted with appropriate utilization of natural resources with support from the government. Through watershed development project there is harmonization of water, soil, forest and pasture resources so as to conserve these resources. In Bankura-1, there are 37 watersheds. Micro Watershed Development is being carried out by NABARD in the scheme NWDPRA. PRADAN has joined the project in 2007. Implementing agency is Panchayat Samiti. PRADAN is acting as Project facilitating agency. PRADAN is working on two fronts—(1) It orients the villagers in water conservation and secondly it provides the technical support. The cost of this support is provided through project management cost. Bankura-1 falls in the region of Agro-ecological zone–7 where central India and part of MP lies. We have already in Bankura there are 4 categories of land here—Toda, Baid, Kanali and Bahal. In tora water holding capacity is low. The land is unterraced. Only pulses and oilseed are grown here. Baid land is sandy loamy in character. Short duration paddy (120 days) like lalat is grown here. The water holding capacity is medium here. Kanali is more loamy and here paddy like lalsarna is grown here. The average productivity is 2-3 tons/he. 50 per cent to 60 per cent land is medium upland, 20 per cent to 30 per cent is low land. Sept is flowering stage of paddy. If there is dry spell, then production of paddy in medium upland is hampered. The water holding capacity of the barren upland is very low. The moisture conservation of the soil is very important. Economic upliftment through direct tangible benefits is very important. In the upland orchard plantation through 30-40 model is being done here.

Check dam is the last option. The approach is 'ridge to valley'. According to catchment area, watersheds can be grouped into various catagories:

(i) Macrowatersheds: 10,000 he
(ii) Microwatersheds: 500-1000 he
(iii) Miliwatersheds: 100-200 he

We did a case study on watershed BD4 Bankura 1 where watershed development has been done. BD3 has four Mouzas. Total area of the watershed is 500 hectores. A total Rs. 30 lakhs has been allocated for the development of watershed.

The Bankura-1 Panchayat Samity has implemented the NABARD sponsored Micro watershed development programme in BD4 watershed area. This watershed programme implemented in BD-4 has following activities.

- 30×40 model with collection pits for soil and moisture conservation: 500 Nos. in 1.1 he of plantation area.
- Forestation consisting of 70,00 timber plants in around 20 he.
- Creation of mango orchard consisting of 4000 plants in around 6 he.
- Water harvesting structure: *Happa* 2 Numbers
- Check Dam–1 No. (Shown in the picture).
- Livelihood support to the landless for livestock rearing–6 groups of about 12 individuals each group have been given support of Rs. 800 per individual.

The expenses for the development of BD-4 are given in Table 5.5.

5.2.1 Impacts

- The soil and moisture conservation of the watershed area has improved.
- Out of 70,000 plants of social forestry (like Akashmoni) 40,000 plants survived. Taking selling price of one plant as Rs. 200, it will fetch Rs. 8 lakhs. Owners of the land will get

Table 5.5: Item-wise Expenditure in BD-4 for Watershed Development

Items	*Expenses (Rs. Lakhs)*	*Percentage*
Social Forestry	7	23.3
Mango Plantation	2.8	9.3
Contour Trenching	1.7	5.7
30-40 model	0.4	1.3
Happa	0.42	1.4
Check Dam	8.5	28.3
Management	4	13.3
Livelihood Support	2.5	8.3
Others	2.68	8.9
Total	30	100.0

Source: Block Development Office, Bankura.

50 per cent and the rest amount will go to the Committee concerned.

- In case of mango orchard the mortality is very high due to bad soil quality (Morrum and rocky), lack of irrigation facility and grazing. Out of 4000 plants, 800 have survived. If each plant gives mangoes of Rs. 200, then the total value of the fruit is Rs. 1,60,000 per year.
- Construction of water harvesting structures like 5 per cent model (Small hupa) have created a strong impact on their livelihood through generation of additional incomes in the area because of the following reasons:
  - ➢ Farmers could provide life saving irrigation to paddy crop during this kharif season;
  - ➢ They could grow vegetables around the bund of hupas.
- Plantation in fallow land and upland has led to check in soil erosion and moisture conservation
- Generation of wage employment opportunity. 1000 mandays have been created for contour trenching. 864 mandays have been created for re-excavation.
- Involvement of local people in the planning and implementation of programme has led to developing more ownership of the programme.
- Active involvement of women SGSY group in all activities has ensured women participation in their village development.

5.2.2 Conclusion

Watershed development is a holistic approach for efficient allocation and development of natural resources. The watershed management programme has strong impact on the livelihoods of rural people. We have visited some other watershed programmes in the Bankura and Purulia also. This has tremendous economic impact for sustainable livelihood generation.

Summary of the Chapter

We have used cost benefit analysis to judge the economic viabability of *happa* (a small private tank) for 9 households. The benefit cost ratio of the projects for all the households is greater than 1. The benefit cost ratio varies from the ratio of 2.46 to 9.69 indicating that the impacts of *Hapa* are not the same across the households. These justify that there is much economic justification for the construction of hapa. We have done a case study of watershed project in Bankura district on the watershed BD-4 in Bankura 1 block. The watershed management programme has strong impact on the livelihood of rural people. We have visited some other watershed programmes in the Bankura and Purulia also. These programmes have tremendous economic impact for sustainable livelihood generation.

Chapter 6

Conclusion and Recommendations

Our analysis reveals that there is huge prospect of tank irrigation in the dry zones of West Bengal in improving the living standard of the people. The cost-benefit analysis result justifies the small decentralized method of tank irrigation. The willingness to pay (WTP) study reveals that farmers are willing to pay for renovation if the water is ensured and ideal institutional mechanism is developed. The profitability analysis shows that there is significant difference across tanks for different crops. So there is scope for enhancing profitabality. The labour demand analysis results show that some crops are more labour-absorbing. But the physical conditions of the most of the tanks surveyed are observed tò be unsatisfactory. Factor analysis results reveal that bad maintenance is the most important component affecting the tank quality. The efficiency analysis of the tanks also shows that there is substantial difference in efficiency levels across tanks. Since Indian agriculture sector will continue to be crucial in providing livelihoods for majority of the population, sustainable irrigation option like tank irrigation should be developed with the highest priority. Block-wise data reveals that there is large amount of unirrigated cultivated land in all the districts and also there is wide gap in the irrigation potential created and the irrigated land. New water harvesting structures need to be created on an urgency basis. Also more attention is required for the improvement of physical conditions and storage capacity of the existing tanks through provisions of more public funds as the existing tanks are much below their capacity with the decline in storage capacity. At the same time, more involvement is required of the beneficiaries for the proper maintenance and better management of the tanks. Innovative experiments like *happa* model may be adopted in these regions as it has already become successful in some parts of these districts. Following are some of the specific recommendations for the improvement of tank irrigation in the dry zones of West Bengal.

1. A number of small tanks over a vast land instead of one big tank at a particular place is more effective. It will prevent water losses and ground water charging will be more. The case study with *Happa* and watershed management in our study area strengthens the point.
2. There is critical need for development of irrigation institution. Irrigators must have incentive for that. In most cases we have observed no existence of water users association or committees (formal or informal) for managing the affairs relating to tanks. Our analysis reveals that higher cropping intensity is associated with the existence of association of tanks. So, WUAs should be formed for each irrigation tank. State laws should be framed in this regard. Panchayats should play greater role in organizing the farmers in forming water users association and assisting them in managing the various problems in tank management. We have observed tank-bed cultivation in some cases. Tank-bed cultivation leads to the deterioration of tanks. Water users association; water committee should monitor these affairs.
3. Tanks and groundwater should not be treated in isolation. Rather an integrated approach should be followed. The recent policy changes in water management have been limited to surface irrigation (canal and tank) and do not include groundwater. Some tanks need to be converted into percolation tanks for recharging ground water, particularly in uplands.
4. Several techniques are now available to minimize evaporation, seepage and percolation losses from water harvesting tanks, which need to be selected for the location-specific application. These techniques should be evaluated and may then be adopted. We have observed no groundwater recharge in our study area. These structures augment water supply and enhance the area under irrigation. There should be more importance on micro irrigation systems like sprinkler and drip irrigation for conservation of water. Despite high

economic benefits of these technologies, the spread is limited because of high initial cost. The methods of cultivation which require less water like SRI should be explored. Efficiency of water use should be improved. The water use efficiency in irrigation will lead to increase in irrigated area. The measures for water use efficiency should be explored further.

5. Imparting orientation programmes on bund maintenance, recharging, rainwater impounding technology and strategies etc. periodically are strongly recommended. Water conveyance channels are earthen made, which breaches frequently. Masonry drains and channels are to be constructed to minimize wastage of water.
6. Siltation can be checked by afforestation. More afforestation programmes need to be taken in the tank bund areas. Sustainable development programme like agroforestry should be practised in the tank catchment area and tank bund. This will check the soil erosion.
7. In the design of the new tanks, local people should be involved. We have observed that some tanks which have been constructed newly are not in proper locations in the sense that water collection capacity of the tank is poor. The direction of the movement of the rain water and the catchment area is very important for constructing new tanks. In constructing and maintaining tanks engineering as well as scientific knowledge should be taken into consideration. We have observed that in some newly constructed tanks grasses have been planted to conserve the soil.
8. In implementing the *Tank Improvement Act* in West Bengal there was Tank Improvement Board in the districts under the District Magistrate who looked after all these affairs. It is unfortunate that this Board does not work for the construction of new tanks. As they are very skill personnel in constructing tanks, they should be engaged in constructing tanks or renovation of tanks.
9. There is urgent need for modernization of old tanks. The bund conditions are very poor in many tanks. For renovation, a lump sum amount of money is required at a point of time. More funds are required for renovation.
10. There is much potential to create year-round income earning opportunities through intensive cultivation of homesteads.

Yet, as there is no irrigation, these lands are very poorly managed. For optimising income from these lands dug wells or small tanks could be created to provide irrigation. Homestead cultivation especially benefits women and offers them an alternative to leaving home in search of wage labour.

11. There is a need to increase the awareness of people, about the water conservation. Water festival may be organized for increasing the awareness about water. Demonstrations of successful renovated projects may be scaled up.
12. In the implementation of the NREGS schemes, there is a restriction of machine use. We have observed that this creates the problem in tank construction or in renovation in the hard rock/dry zone areas. There should be flexibility in laws in this regard or other programmes should be converged with NREGS. There should be proper counselling by which households may be motivated to apply for renovation works in their own tanks through NREGS. We have observed that some households are reluctant to use NREGS funds for renovation works on the fear that this might lead to losing their right on the tank.
13. In some places we have observed that there is high demand for *'Indara'* which is big dug well. One *indara* is sufficient for irrigation need of one household. The farmers prefer it because its longevity is high and the water holding capacity is also good. But its construction cost is high. The tube well irrigation should be made complementary to tank irrigation depending upon the groundwater availability.
14. We have observed that in some cases the owner or the tank lives far off from the tank leading to deterioration of the tank. A viable negotiation with the owner and the local people is warranted.
15. In some areas NGOs like PRADAN are doing good jobs in irrigation management through local capacity building and imparting skills to the farmers. Government should try to make partnership with these types of agents.
16. There is huge scope of aquaculture in these bundhs/tanks. But because of lack of understanding among the villagers, aquaculture is not being developed. Efforts should be made to increase pisciculture in these tanks. A certain amount of sale proceeds of fish may be distributed among the villagers.

17. Some households could not use tank water because they do not have enough purchasing power to access pumps and pipe. Government should make provisions for easy loans for purchasing the equipments.
18. The details of all irrigation tanks should be with panchayat office. We have observed that in most of the GP offices records of the tanks in regard to exact location (mapping), area, depth, present condition etc. are not kept. The GP should take initiative to prepare the resource maps.
19. Irrigation department is the agent of change. As we have observed in some cases, the attitude of irrigation department as providers should be changed. We have also observed that the link between the irrigation department and the local people is weak.
20. We have observed from the household survey that extension services are very much lacking in the rural areas. Extension personnel should be properly trained relating to the type of crop to be cultivated and irrigation efficiency measures.
21. Keeping in view the direction of the gravity of water flows and terrain, watersheds should be developed. As our study reveals watershed development with tank irrigation has huge impact on agricultural productivity and rural livelihood.
22. User charges for use of tanks may be introduced for ensuring financial sustainability of the tanks.
23. Follow-up action is required to ensure equality in water distribution.
24. Rights to water should be given also to the landless.
25. Rural people lack technical know-how to prepare a plan for Integrated Natural Resource Management (INRM) of the rural community especially in an area with complex ecological scenario. They also lack the orientation of how to implement the programmes at the grassroot level by involving the entire villagers. They should properly be trained. A new generation of local leaders need to be created through systematic training.
26. Massive capacity building exercises are required for government agencies, NGOs and user groups. Local institutions for managing waterbodies whether formal or informal should be strengthened. A manual for tank development may be prepared.

27. We have observed that link between different government line departments are very poor in respect of construction or re-excavation of tanks. For development of tank irrigation, there should be close co-ordination among different line departments like irrigation, agriculture and soil departments.
28. Crop Sequence is very important concept for conservation of soil and water. As recommended by the Dryland Research Centre following sequence of crops should be followed for Red and Laterite zone of West Bengal:
    a. In upland (it is called tar situation), where rice cannot be grown successfully, crops like, Arhar, maize, cowpea, soybean, kalai and groundnuts are suitable for cultivation during *kharif* season.
    b. In medium lands (*Baid* situation), besides the above crops direct seeded short duration HYV rice varieties like, Aditya, Heera, Pravat.
    c. Medium land (it is called *Kanali*) - *Lalat* rice variety (Medium duration: 120 days) is very popular in Bankura region.
    d. Low land (*Sol* situation)—*Swarna* rice variety (Long duration paddy variety: 135-140 days) *Kharif* season: Maize/Cowpea, Post *Kharif* season: Niger, Kulti, Sesame, Toria.
    e. Inter-cropping can be also recommended for improving living standard of farming community and sustainable crop production.

References

ADB (2006), *Rehabilitation and Management of Tanks in India*, Asian Development Bank, http://www.adb.org/documents/studies/tanks-india/rehabiliation-management-tanks.pdf

Agarwal, Anil (2001), "Water Harvesting in a New Age", in Anil Agarwal *et al.* (eds.), *Making Water Everybody's Business*, Centre for Science and Environment, New Delhi.

Agarwal, Anil *et al.* (eds.) (2001), *Making Water Everybody's Business*, Centre for Science and Environment, New Delhi.

Ahmed, Akhter U. and Rajan, K. Sampath (1992), "Effects of Irrigation-Induced Technological Change in Bangladesh Rice Production", *American Journal of Agricultural Economics,* Vol. 74, No. 1, February, pp. 144-157.

Ahuja, A. (2006), *Agriculture and Rural Development in India*, New Century Publications.

Ali, N. (2008), "Rice Yield Gap in West Bengal", *Agricultural Situation in India*, March 2008.

Anbumozhi, V., K. Matsumoto and E. Yamaji (2007), "Sustaining Agriculture through Modernization of Irrigation Tanks: An Opportunity and Challenge for Tamil Nadu, India". http://cigr-ejournal.tamu.edu/submissions/volume3/LW%2001%20002.pdf.

Anderson, Tim (2007), A Data Envelopment Analysis (DEA) Home Page *http://www.etm.pdx.edu/dea/homedea.html#DEA_Home_Page_ABOUT*

Arya, Swarn Lata, R.C. Kaushal and S.S. Grewal (1994), "Economic Viability of Watershed Management Project Selected to Rehabilitated Degraded Aravali Foothills of Haryana", *Indian Journal of Agricultural Economics*, Vol. XLIX, Oct.-Dec., pp. 591-600.

Ashturkar, B.W. (1986), "Irrigation and Water Management", *Indian Journal of Agricultural Economics*, Vol. XLI, Oct.-Dec., pp. 523-528.

Asian, Development Bank (2008), *Irrigation Management Transfer—Strategies and Best Practices,* Sage Publications.

Babu, B.V. and V. Ramakrishna (2008), "Water Crisis: Origin, Identification and Management", in Gupta K.R. ed.

Bagi, F.S. (1981), "Economics of Irrigation in Crop Production in Haryana", *Indian Journal of Agricultural Economics*, Vol. 36, Jul-Sep, pp. 15-26.

Balasubramaniank, R. and N. Selvaraj (2003), "Poverty, Private Property and Common Pool Resource Management: The Case of Irrigation Tanks in South India", Working Paper No. 2-03, *SANDEE.*

Banda, B.M. *et al.* (2007), "Determinants of Quality and Quantity Values of Water for Domestic Uses in Steelpoort Sub-basin: A contingent Valuation Approach", in Perret *et al.* eds.

Banik, Arindam (1994), "Technical Efficiency of Irrigated Farms in Villages of Bangladesh", *Indian Journal of Agricultural Economics*, Vol. XLIX, Jan.-Mar., pp. 70-78.

Bardhan, Pranab (2000), "Irrigation and Cooperation: An Empirical Analysis of 48 Irrigation Communities in South India", *Economic Development and Cultural Change*, Vol. 48, No. 4, July, pp. 847-865.

Bardhan, Pranab K. (1979), "Agricultural Development and Land Tenancy in a Peasant Economy: A Theoretical and Empirical Analysis", *American Journal of Agricultural Economics*, Vol. 61, No. 1, Feb., pp. 48-57.

Basak, N.N., (2008), *Irrigation Engineering*, Tata McGraw-Hill Publishing Company Limited, New Delhi.

Basu, Subhashranjan eds. (2002), *Changing Environmental Scenario of the Indian Sub-Continent*, ACB Publications, 2002

Bhattarai, M. Anand (2009), "Towards a legal Framework", in Anil Agarwal *et al.* eds., *Making Water Everybody's Business*, Centre for Science and Environment, New Delhi.

Bhende, M.J. (2007), "Conflicts over Tank-bed Cultivation in Karnataka: Command Area Farmers and Official Encroachers", in Joy, K.J. *et al.* eds. (2007).

Bhullar, A.S. *et al.* (2006), "Integrated Land and Water Use A Case Study of Punjab", *Economic and Political Weekly*, December 30, pp. 5353-5357.

Biswas A.K. *et al.* (1997), *National Water Master Plans for Developing Countries*, Oxford University Press. New Delhi

Central Water Commission (CWC), *Handbook of Water Resource Statistics*, http://cwc.gov.in/main/webpages/publications.html

Chakravorty, K.K. *et al.* eds. (2006), *Traditional Water Management Systems of India,* Aryan Books International, New Delhi.

Chakravorty Ujjayant and James Roumasset (1991), "Efficient Spatial Allocation of Irrigation Water", *American Journal of Agricultural Economics*, Vol. 73, No. 1, Feb., pp. 165-173.

Chandrakanth, M.G., and V. Arun (1997), "Externalities in Ground Water Irrigation in Hard Rock Areas", *Indian Journal of Agricultural Economics*, Vol. 52, No. 4, Oct-Dec.

Chatterjee, Nandini (1995), *Irrigated Agriculture*, Rawat Publications, Jaipur.

Chattopadhyaya, B.D. (1973), "Irrigation in Early Medieval Rajasthan", *Journal of the Economic and Social History of the Orient*, Vol. 16, No. 2/3, Dec., pp. 298-316.

Chidnanda, B.L., Gracy. P.C. and Suryaprakash, S. (1996), "Role of Community in Irrigation Management: Sustainability of Traditional Water Harvesting Practices in South India", University of California.

Chiranjeevulu, P. (1992), *Tank Irrigation and Agricultural Development*, Kanishka Publishing House, Delhi.

Chopra, K. (2003), "Sustainable Use of Water—The Next Two Decades", *EPW*, pp. 3360-3365.

Chopra, Kanchan (1989), "Water Resource Management", *Indian Journal of Agricultural Economics*, Vol. XLIV, Oct.-Dec., pp. 404-406.

Chopra, Kanchan and Gopal Kadekodi (1993), "Watershed Development, A Contrast with NREP/JRY", *EPW*.

CII, (2007), Sustainable Economic Development—A Perspective, in.kpmg.com

Coelli, Tim (1996), A Data Envelopment Analysis Computer Programme, http://www.une.edu.au/econometrics/cepa.htm.

Cohen, Jerome B. (1953), "Economic Development in India", *Political Science Quarterly*, Vol. 68, No. 3, Sep., pp. 376-395.

Cornish, G. with B. Bosworth, C. Perry, J. Burke (2004), "Water Charging in Irrigated Agriculture— Analysis of Interrnational Experience", *Food And Agriculture Organization of the United Nations*, Rome.

Desarda, M.H. (2009), "Towards Community-based Rainwater Harvesting: Some Conceptual and Policy Issues", in Anil Agarwal *et al.* eds., *Making Water Everybody's Business*, Centre for Science and Environment, New Delhi.

Despande, R.S. (2008), "Watersheds: Putting the Cart before the Horse", *EPW*, February 9, pp. 74-76.

Dhawan, (1989), "Major and Minor Irrigation Works", *EPW*, September 30, pp. A117-A123.

Dhawan, B.D. (1985), "Questionable Conception and Simplistic Views about Irrigated Agriculture of India", *Indian Journal of Agricultural Economics*, Vol. XL, Jan-March, pp. 1-13.

Dhawan, B.D. (1986), "Irrigation and Water Management in India", *Indian Journal of Agricultural Economics*, Vol. XLI, Oct.-Dec., pp. 271-281.

Dhawan, B.D. (1989), "Water Resource Management: Issues and Dimensions", *Indian Journal of Agricultural Economics*, Vol. XLIV, Oct.-Dec., pp. 233.

Dhawan, B.D. (1977), "Tube well Irrigation in the Gangetic Plains", *EPW*, September 1977.

Dhawan, B.D. (1985), "Irrigation Impact on Farm Economy", *EPW*, September.

Dhawan, B.D. (1988), "Indian Irrigation—An Assessment", *EPW*, May 7, pp. 965-972.

Droogers, Peter and Geoff Kite, "Estimating Productivity of Water at Different Spatial Scales Using Simulation Modeling", *International Water Management Institute*, PO Box 2075, Colombo, Sri Lanka.

Dutta, M. *et al.* (2005), "Economics of Conservation of Built Heritage—A CV

Study of Some Heritage Sited of Culcutta", *Indian Economic Review,* Vol. XXXX, No. 2, 2005, pp. 221-243.

Easter, K. William (1977), "Improving Village Irrigation Systems: An Example from India", *Land Economics*, Vol. 53, No. 1, Feb., pp. 56-66.

Easter, K. William (2000), "Asia's Irrigation Management in Transition: A Paradigm Shift Faces High Transaction Costs", *Review of Agricultural Economics*, Vol. 22, No. 2, Autumn - Winter, pp. 370-388.

Easter, K. William and Martin E. Abel, George Norton (1977), "Regional Differences in Agricultural Productivity in Selected Areas of India", *American Journal of Agricultural Economics*, Vol. 59, No. 2, May, pp. 257-265.

Fried, H.O. *et al.* (1993), *The Measurement of Productive Efficiency*, Oxford University Press, USA.

Gang, L. and B. Flemingham (2002), *The Technical Efficiency of Australian Irrigation Schemes*, School of Economics, University of Tasmania, available in the website.

Gangwar, A.C. and W.H. Vanden Toorn (1987), "The Economic of Adverse Groundwater Condition in Haryana State", *Indian Journal of Agricultural Economics*, Vol. XLII, April-June, pp. 160-172.

George, P.S. (1999), "Some Reflections on Food Security in India", *Indian Journal of Agricultural Economics*, Vol. 54, No. 4, pp. 465-489.

Gireesh, M., N. Nagaraj and M.G. Chandrakanth (1997), "Rehabilitation of Irrigation Tanks in Eastern Zone of Karnataka—An Economic Analysis", *Indian Journal of Agricultural Economics*, Vol. 52, April-June, pp. 231-243.

Gole, Prakash (2009), "Biodiversity in Artificial Wetlands", in Anil Agarwal *et al.* (eds.), *Making Water Everybody's Business*, Centre for Science and Environment, New Delhi.

Gosh, P.K. (1987), "Effect of Surface Irrigation on Land Degradation—Problems and Strategies", *Indian Journal of Agricultural Economics*, Vol. XLII, July-Sep., pp. 416-423.

Goswami, S. (2006), "Tank Irrigation in West Bengal: An Overview", *ICFAI Journal of Agricultural Economics*, Vol. 3, Issue 4, pp. 65-73.

Gould, John (2009), "Health and Rainwater Potability, in Anil Agarwal *et al.* (eds.), *Making Water Everybody's Business*, Centre for Science and Environment, New Delhi.

Government of West Bengal (2006), *Estimates of Area and Production of Principal Crops in 2004-05*, Directorate of Agriculture.

Government of India (2006), *Seasonal Variation in the Operational Land Holdings in India*, 59th Round NSSO.

Government of India (2006), *Seasonal Variations in the Operational Land Holdings of India 2002-03*, NSSO 59th Round, Ministry of Statistics and Programme Implementation.

Government of India, *Economic Survey*, Various Issues.

Government of West Bengal (2003), *WIDD, Report on Third Minor Irrigation Census in West Bengal (2000-01).*

Government of West Bengal (1967), *The Bengal Tanks Improvement Act, 1939* (As modified upto the 1st April, 1967).

Government of West Bengal (2003), *The Statistical Abstract*, Bureau of Applied Economics and Statistics.

Government of West Bengal (2008), *Approach to the Eleventh Five-Year Plan for West Bengal*, Development and Planning Department,

Government of West Bengal, *Statistical Abstract*, Various years. Bureau of Applied Economics and Statistics.

Government of West Bengal, *Economic Review*, Various Years.

Gulati, A. *et al.* (2006), *Institutional Reforms in Indian Irrigation*, Sage Publications, New Delhi.

Gupta, Dipankar (1986), "Tribal Development in West Bengal District", *EPW*, January 4.

Gupta, K.R. (2008), *Water Crisis in India*, Atliantic Publishers and Distributors (P) Ltd., New Delhi.

Gurjar, R.K. and B.C. Jat (2008), *Geography of Water Resources*, Rawat Publications, New Delhi.

Hill, R.C., William E Griffiths and G.C. Lim (2008), *Principles of Econometrics*, John Wiley and Sons, New York

Hooja, R. *et al.* (2007), *Water Management—Multiple Dimensions*, Rawat Publications, New Delhi.

ICEF (1999-2006), *Rehabilitation of Integrated Tank Management System in the Kaliveli Watershed Villupuram District, Tamil Nadu*", *Centre for Ecological Landuse*, Water Management and Rural Development, Auroville.

Imam, Shahid (2006), "Economic Impact of Indira Gandhi Canal Project in Western Rajasthan: A Village Level Study", *Indian Journal of Regional Science*, Vol. XXXVIII.

Iqtidar, Husain Siddiqui, (1986), "Water Works and Irrigation System in India during Pre-Mughal Times", *Journal of the Economic and Social History of the Orient*, Vol. 29, No.1, Feb., pp. 52-77.

Iyer, R.R. (2008), "Water: A Critique of Three Concepts", *EPW*, January 5, pp. 15-18.

Iyer, Ramaswamy R. (2009), "Wanted: Fresh Ideas", in Anil Agarwal *et al.* (eds.), *Making Water Everybody's Business*, Centre for Science and Environment, New Delhi.

Jana, Sebak Kumar (2010), "Prospect of Sustainable Tank Irrigation Options in the Dry Zones of West Bengal", Research Project Sponsored by and submitted to ICSSR, New Delhi.

Jana, Sebak Kumar (2011), "Sustainable Small Scale Irrigation in the Dry Zones: A Case Study on *Happa* (Small Tank) Model in the State of West Bengal, India", Paper Presented in the International Conference on "*Innovation and Sustainability Transitions in Asia*" at University of Malaya, Malaysia, 9-11 January 2011, Published in D.N. Bhattacharyya (ed.) "*Indian Economic Development—Contemporary Issues, A Festschrift in Honour of Professor Raj Kumar Sen*", Regal Publications, New Delhi.

Jana, Sebak Kumar, "Willingness to Pay for Tank Maintenance: A Study on Tank Irrigation in the Dry Zones of West Bengal in India", *International Journal of Research in Commerce*, Economics and Management (Accepted for Publication).

Jana, Sebak Kumar, K. Palanisami and Amit Das (2012), "A Study on Tank Irrigation in the Dry Zones of West Bengal", *Indian Journal of Agricultural Economics*, Vol. 67, No. 2, April to June.

Janakarajan, S. (1993), "In Search of Tanks", *EPW*, June 26, pp. A 53- A60.

Joirath, J. (1985), "Technical and Institutional Factors in Utilisation of Irrigation—A Case Study of Public Canals in Punjab", *EPW*, March 1985.

Joshi, Deepa (2004), *Secure Water—Whither Poverty? Livelihoods in the DRA: A Case Study of the Water Supply Programme in India*. Overseas Development Institute.

Joy, K.J. *et al.* eds. (2007), *Water Conflicts in India*, Centre for World Solidarity.

Kakumanu, Krishna Reddy, Siegfried Bauer, T.S. Amjath Babu (2006), "Participatory Irrigation Management through Moral Sausion: A Case Study Comparing the Formal and Informal Organisational Structures", *Conference on International Agricultural Research for Development*, University of Bonn, Oct.

Kalirajan, K. and T.C. Flinn (1981), "Allocative Efficiency and Supply Response in Irrigated Rice Production", *Indian Journal of Agricultural Economics*, Vol. 36, April-June, pp. 16-24.

Kanagaraj, (2005), "Performance of Irrigation Tanks in Tamil Nadu", *Agricultural Studies in India*, November 2002.

Kannan, K.P. (1989), "Agricultural Stagnation and Irrigation", *EPW*, May 13, pp. 1067-1068.

Kashwan, Prakash (2007), "Traditional Water Harvesting Structure: The Community behind Community" in Joy, K.J. *et al.* eds.

Khanka, S.S. *et al.* (2006), "Participatory Watershed Management", *Kurukshetra*, July, pp. 12-16.

Kulkarni, P. (2003), "The Sinking Lifts: Government Irrigation Schemes in North Karnataka", *EPW*, pp. 3968-3972.

Kumar, M.D. (2007), "Assessing Economic Value of Ground Water—Illustrations from North Gujarat", in Kumar M.D. ed.

Kumar, M.D. (2007), *Groundwater Management in India*, Sage Publications, New Delhi.

Lahiri-Dutta, Kuntala and Robert J. Wasson (2008), *Water First: Issues and Challenges Nations and Communities in South Asia*, Sage Publications, New Delhi.

Madhav, Roopa, *Irrigation Reforms in Andhra Pradesh*, International Environmental Law Research Centre, *info@ielrc.org, www.ielrc.org.*

Majumdar, D.K. (2006), *Irrigation Water Management—Principles and Practice*, Prentice Hall of India.

Maloney, C. and K.V. Raju (1994), *Managing Irrigation Together—Practice and Policy in India*, Sage Publications. New Delhi.

Mitra, Ashok K. (1984), "Managing Irrigation System in Drought-Prone Area", *Indian Journal of Agricultural Economics*, Vol. XXXIX, July-Sep. pp. 491-505.

Mitra, Ashok K. (1992), Joint Management of Irrigation Systems in India—Relevance of Japanese Irrigation, *EPW*, June 27.

Mittal, S.P. (2005), "Rainwater Management in Shibalik Region", *Kurukshetra*, pp. 16-19. in Verma *et al.* eds. (2009).

Money, R.I. (1919), "The Irrigation of the Konia Plain", *The Geographical Journal*, Vol. 54, No. 5, Nov., pp. 298-303.

Mosse, David (1999), "Colonial and Contemporary Ideologies of 'Community Management': The Case of Tank Irrigation Development in South India", *Modern Asian Studies*, Vol. 33, No. 2, May, pp. 303-338.

Nagraj, N. and M.G. Chandrakanth (1995), "Low Yielding Irrigation Wells in Penninsular India—An Economic Analysis", *Indian Journal of Agricultural Economics*, Vol. L, Jan-March, pp. 47-58.

Narasaiah, M.L. (2003), *Food Security and Irrigation*, Discovery Publishing House, New Delhi.

Narasimha, Rao P. (2007), *Irrigation Development—Issues and Challenges*, Discovery Publishing House.

Narayanamoorthy, A. and R.S. Deshpande (2003), "Irrigation Development and Agricultural Wages", *EPW*, August 30, pp. 3716-3722.

Narayanamoorthy, A. and R.S. Deshpande (2005), *Where Water Seeps! Towards a New Phase in India's Irrigation Reforms,* Academic Foundation, New Delhi.

Osman, M. *et al.* (2007), "Rejuvenating Tanks for Self Sustainable Rainfed Agriculture in India", pp. 67-70.

Pain, Asis Kumar and Samrat Goswami (eds.) (2008), *Participatory Irrigation Management: Issues and Country Experiences*, IUP, Agartala.

Palanasami, K. and K. William (1984), "Irrigation Tanks of South India: Management Strategies and Investment Alternatives", *Indian Journal of Agricultural Economics*, Vol. XXXIX, April-June, pp. 214-223.

Palanisami, K. and K.W. Easter (2000), *Tank Irrigation in the 21st Century*, Discovery Publishing House.

Palanisami, K. and Balasubramanian, B.R. (2009) "Reconciling the Private Well with the Community Tank", in Anil Agarwal *et al.* (eds.), *Making Water Everybody's Business*, Centre for Science and Environment, New Delhi.

Palanisami, K. (2007), "Sustainable Management of Tank Irrigation Systems in India", http://www.nourin.tsukuba.ac.jp/~tasae/2005/India.pdf

Pant, Niranjan (1998), "Indigenous Irrigation in South Bihar", *EPW*, December 5.

Parikh, Ashok and Prabin Trivedi (1982), "Impact of Irrigation and Fertilizers on the Growth of Output in A.P.: An Bayesian Approach", *Indian Journal of Agricultural Economics*, Vol. 37, April-June, pp. 159-170.

Pasricha, N.S. (2002), "Water Resources and Management for Sustained Crop Production in Punjab", October.

Pasricha, N.S. (2008), "Optimising Water Resources Management for Sustainabe Use", *Agricultural Situation in India,* Vol. 64, Issue 11, pp. 547-554.

Pretty, J.N. and A.S. Ball, Li Xiaoyun, N.H. Ravindranath (2002), "The Role of Sustainable Agriculture and Renewable-Resource Management in Reducing Greenhouse-Gas Emissions and Increasing Sinks in China and India", *Philosophical Transactions: Mathematical, Physical and Engineering Sciences*, Vol. 360, No. 1797.

Punmia, B.C. and Pande, B.B. Lal (2007), *Irrigation and Water Power Engineering*, Laxmi Publications Pvt. Ltd., New Delhi.

R.H. Whitbeck (Oct., 1988), "*Irrigation in the United States*", the *Geographical Journal*, Vol. 54, No. 4, (Oct., 1919), pp. 221-231.

Raghunath, T.P. and V. Vasanthan (2007), "The Bahoor Cluster of Irrigation Tanks in Pondicherry— Usufruct Sharing between the Dalit Hamlet and the Main Village", in Joy K.J. *et al.* (eds.).

Raju, K.V. (2007), "Micro-Level Water Conflicts—A Review", in Joy *et al.* (eds.).

Raju, K.V. (2001), Participatory Irrigation Management in India (Andhra Pradesh). Accessed at ftp:/ftp.fao.org/agl/aglw/imt/csandna_pradesh.pdf.

Ramanathan, R. (2003), *An Introduction to Data Envelopment Analysis*, Sage Publications. New Delhi.

Rao, C.H. Hanumantha, (2005), *Agriculture, Food Security, Poverty and Environment*, Oxford University Press. New Delhi.

Rao, V.M. and M.G. Chandrakant (1984), "Resources at the Margin—Tank Irrigation Programme in Karnataka", *EPW*, Vol. XIX, No. 26, June 30.

Rao, V.M. (1978), "Linking Irrigation with Development—Some Policy Issues", *EPW*, June 17.

Reddy, K. Ramakrishna (1995), *Irrigation and Agricultural Development in India*, Ashish Publishing House, New Delhi.

Reddy, Krishna *et al.* (2008), "Participatory Irrigation Management Through Moral Suastion". In Pain *et al.* eds., IUP.

Reddy, P.S. (1997), *Different Sources of Irrigation*, Manak Publications, New Delhi.

Reddy, V.R. (2004), "Collective Action on Water", *EPW*, July 17.

Reddy, V.R. (2006), "Jalayagnam and Bridging Regional Disparities—Irrigation Development and Distribution in Andhra Pradesh", *EPW*, November 4.

Reddy, V.R. (2009), *Water Security and Management—Ecological Imperatives and Policy Options*, Academic Foundation, New Delhi.

Reddy, V.R. and S. Mahendra Dev (2006), *Managing Water Resources—Policies, Institutions and Technologies,* Oxford University Press. New Delhi.

Reddy, V.R. *et al.* (2005), "How Participatory is Participatory Irrigation Management—Water Users Associations in Andhra Pradesh?" *Economic and Political Weekly*, December 31, pp. 5587-5595.

Reddy, V.R. (1990), "Irrigation in Colonial India—A Study of Madras Presidency During 1860-1900", *EPW*, May 5-12.

Reddy, V.R. (2002), "Water Security and Management—Lessons from South Africa", *EPW*, July 13, pp. 2878-2881.

Rockstrom, J. (2005), "Managing Rain for the Future", in Figures *et al.* (ed.) (2005), *Rethinking Water Managemen—Innovative Approaches to Contemporary Issues*, Earthscan.

Roy, S. (ed.), (2006), *Oxford Handbook of Agriculture*, Oxford University Press. New Delhi.

Rydzewski, Janusz, R. (1990), "Irrigation: A Viable Development Strategy?", *The Geographical Journal*, Vol. 156, No. 2, Jul., pp. 175-180.

Sakhtivadivel, R. *et al.* (2004), "Rejuvenating Irrigation Tanks through Institutions", *EPW*, July 31.

Sampath, Rajan K. (1990), "On Some Aspects of Irrigation Distribution in India", *Land Economics*, Vol. 66, No. 4, Nov., pp. 448-463.

Sangal, S.P. (1991), "Pricing of Irrigation Waters in India", *EPW*, November 16.

Sangma, A.P. (2009), "Recognising the Inevitable", in Anil Agarwal *et al.* eds., *Making Water Everybody's Business*, Centre for Science and Environment, New Delhi.

Sastry, G. *et al.* (2003), "Reshaping of Dry Land Agriculture Through Watershed Management Programmes in Different AERs of India", *Agricultural Situation in India*, November, pp. 503-510.

Sastry, G. *et al.* (2004), "Watershed Management Programmes--Physical and Socio-Economic Impacts in Semi-Arid Regions of India", *Agricultural Studies in India*, Sept.

Satyasai, K.J.S. and K.U. Viswanathan (1997), "Evaluation of Alternative Water Management Strategies for Water Scarce Areas", *Indian Journal of Agricultural Economics,* Vol. 52, July-Sep., pp. 499-507.

Sekar, I. and C. Ramasami (2008), "Econometric Analysis of Farmers' Participation in Tank Modernisation", *Agricultural Situation in India*, February, pp. 555-562.

Selvarajan, S. and S.R. Subramaniam (1981), "Economics Impacts of Resource Use Optimization and Water Augmentation in Farms of Parambikulam Aliyar Project Regions", *Indian Journal of Agricultural Economics*, Vol. 36, Jan.-Mar., pp. 89-100.

Sengupta, Nirmal (1991), *Managing Common Property—Irrigation in India and Philippines*, Sage Publications.

Sengupta, Nirmal (1985), "Irrigation-Traditional Vs. Modern", *EPW*, November, pp. 1919-1938.

Shah, Amita (2005), "Economic Rationale, Subsidy and Cost Sharing in Watershed Projects", *Economic and Political Weekly*, June 25, pp. 2663-2671.

Shah, Esha (2003), *Social Designs: Tank irrigation Technology and Agrarian Transformation in Karnataka, South India*, Oriented Longman Private Limited, New Delhi.

Shah, Esha (2007), "Head Versus Tail—Conflict in Tank Command in Karnataka", In Joy K.J. *et al.* (eds.).

Shah, Mihir (2009), "Participarory Watershed Development: An Institutional Framework", in Anil Agarwal *et al.* eds., *Making Water Everybody's Business*, Centre for Science and Environment, New Delhi.

Shah, T. *et al.* (2006), "Is India Ripe for Integrated Water Resources Management? Fitting Water Policy to National Development Context", *Economic and Political Weekly*, August 5, pp. 3413-3421.

Shaheen, F.A. and R.L. Shiyani (2005), "Equity Redistribution: Groundwater Bore-wells in North Gujarat", EPW, Jan 22.

Sharma, Rita (2002), "Reforms in Agricultural Extension", *Economic and Political Weekly*, June 27—August 2, pp. 3124-3331.

Sidhu, D.S. (1984), "An Economic Analysis of Various Source of Irrigation in Panjab", *Indian Journal of Agricultural Economics*, Vol. XXXIX, July-Sep., pp. 506-520.

Singh, Anil (2009), "Wanted: People-oriented Research", in Anil Agarwal *et al.* eds., *Making Water Everybody's Business*, Centre for Science and Environment, New Delhi.

Singh, Rajender (1994), Social Cost Benefit Analysis of Investment in Irrigation, *EPW*, May 14.

Sivasubramaniyan, K. (1994), Decline of Tank Irrigation, *EPW,* February 26.

Sivasubramaniyan, K. (2006), "Sustainable Development of Small Water Bodies in Tamil Nadu", *EPW*, June 30.

Sreedhar, G. (2007), "Tank Irrigation Management at the Peril: Emerging Issues and Challenges" in P. Narasimha Rao (ed.).

Swaminathan, M.S. (2001), "Science and Our Agricultural Future", *Agricultural Studies in India*, pp. 175-185.

Thakkar, H. (1999), "Assessment of Irrigation in India", Contributing Paper to the World Commission on Dams, *http:/www.dams.org*

Thakur, Jawahar and Praduman Kumar (1984), "A Comparative Study of Economic Efficiency of Different Irrigation System in Western Uttar Pradesh", *Indian Journal of Agricultural Economics*, Vol. XXXIX, July-Sep., pp. 521-527.

Tsur, Yacov and Ariel Dinar (1997), "The Relative Efficiency and Implementation Costs of Alternative Methods for Pricing Irrigation Water", *The World Bank Economic Review*, Vol. 11, No. 2, May, pp. 243-262.

Vaidyanathan, A. (2006), *India's Water Resources, Contemporary Issues of Irrigation*, Oxford India Paperbacks, New Delhi.

Vaidyanathan, A. (2009), "A Case for Rainwater Harvesting", in Anil Agarwal *et al.* eds., *Making Water Everybody's Business*, Centre for Science and Environment, New Delhi.

Vakulabharanam, V. (2004), "Agricultural Growth and Irrigation in Telangana", *EPW*, March 27, pp. 1421-1426.

Vani, S.M. (2009), "Enabling Law for Water Harvesting", in Anil Agarwal *et al.*

(eds.), *Making Water Everybody's Business*, Centre for Science and Environment, New Delhi.

Vasimalai, M.P. (2006), "Shaping Stakeholders' Perspective: Survival of the Common Traditional Tanks in South India", Accessed at dlc.dlib.indian/dic/handle/10535/1760.

West Bengal Pollution Control Board (WBPCB) (2009), *Water Resource and Quality in West Bengal—A State of Environment Report*, Jayasree Press, Kolkata.

William, K. Easter (1977), "Improving Village Irrigation System: An example from India", University of Wisconsin Press http://www.Jstor.org.

Index